FOUNDATIONS AND APPLICATIONS OF GRAPH THEORY

FROM BASICS TO ADVANCED CONCEPTS

DR. RAJPAL KOSALIYA

Associate Professor and Head of Department,
School of Basic and Applied Sciences,
Raffles University, Neemrana

Title   :   Foundations and Application of Graph Theory : From Basics to Advanced Concepts

Author   :   Dr. Rajpal Kosaliya

Edition   :   First (December, 2024)

ISBN   :   9789348037237

Published by

TANEESHA PUBLISHERS

A Venture by -
PRACHI DIGITAL PUBLICATION

Regd. Add.: 254, Khuriyakhatta No. 10, Bindukhatta, Lalkuan, Nainital - 262402, Uttarakhand, India

Website : www.taneeshapublishers.in

E-mail : taneeshapublishers@gmail.com

Phone : +91 8454 812712, +91 8057 812712

Printed by :

Manipal Technologies Limited, Bengaluru - 560001, Karnataka

PREFACE

Graph theory, an intriguing and vital area of mathematics, finds its roots in the 18th century, with Leonhard Euler's pioneering work on the Seven Bridges of Königsberg problem. Since then, the field has burgeoned into a versatile and indispensable tool, influencing diverse areas like computer science, biology, sociology, and network theory. The foundational structures and profound applications of graph theory underscore its role as a linchpin in solving modern-day problems.

This book, "Foundations and Applications of Graph Theory: From Basics to Advanced Concepts," is born out of a passion for unraveling the elegance and utility of graphs. It aims to bridge the gap between theoretical foundations and practical applications, catering to students, researchers, and professionals alike. With its structured progression from introductory topics to advanced themes, the book serves as both an educational guide and a reference for exploring the depths of graph theory.

Through its chapters, readers will journey from the rudiments of graph definitions and properties to specialized topics like magic labeling, graph coloring, and isomorphism. The exploration is not merely theoretical; real-world applications across disciplines are woven throughout to demonstrate the relevance and adaptability of graph theory.

I extend my heartfelt gratitude to my mentors, colleagues, and students whose insights and queries have enriched my understanding and teaching of this subject. Their encouragement has been instrumental in shaping this work. A special thanks to Raffles University, Neemrana, for fostering an environment of academic growth and collaboration.

It is my hope that this book inspires a deeper appreciation for the beauty of graph theory and equips its readers to leverage its concepts to innovate, analyze, and solve problems in their respective fields. Any feedback, suggestions, or discussions from readers will be warmly welcomed and appreciated, as ay are vital for continuous learning and improvement.

Dr. Rajpal Kosaliya

CONTENTS

CHAPTER 1

INTRODUCTION TO GRAPHS

1.1 BASIC DEFINITIONS AND TYPES OF GRAPHS

Definition of a Graph

A graph, formally denoted as $G = (V, E)$, is a mathematical composition representing relationships between objects. It contains of two sets:

- **V**: The set of **vertices** (also called nodes), representing the objects or entities in the graph.
- **E**: The set of **edges**, representing the connections or relationships between the vertices.

Edges can be **directed** or **undirected**, depending on whether the relationship between vertices is one-way or two-way.

Real-world Applications:

Graphs find applications in various domains, modeling real-world scenarios:

- **Social Networks**: Vertices represent individuals, and edges represent bonds or connections.
- **Transportation Networks**: Vertices represent cities or places, and edges represent roads or transportation routes.
- **Computer Networks**: Vertices represent computers or devices, and edges represent communication links.
- **Project Management**: Vertices represent tasks, and edges represent dependencies between tasks.

Types of Graphs

1. Simple Graph

A **simple graph** is where:

- **No loops:** There are no edges connecting a vertex to itself.
- **At most one edge:** There is at most one edge connecting any two separate vertices.

Example:

A social network where each individual is connected forir friends with a single line representing their friendship, without self-connections.

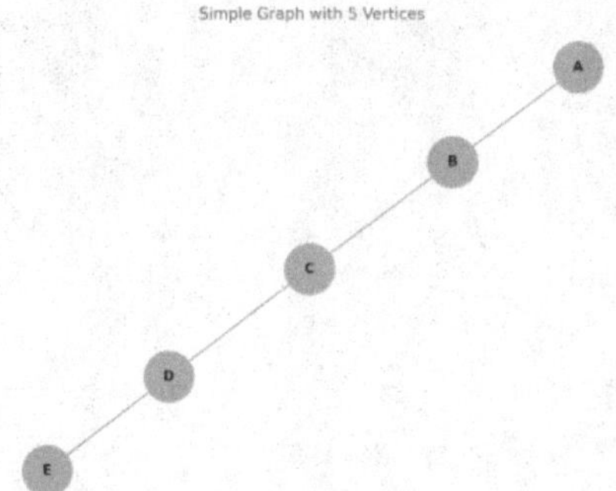

Figure 1.1 | Simple Graph Representation of 5 Vertices

Vertices (Nodes):

A, B, C, D, E represent individual entities (e.g., people, locations, or objects).

Edges (Lines):

Connections represent relationships or links between adjacent nodes.

Edges in this graph are **undirected**, indicating mutual relationships between connected nodes.

Properties:

- **Simplicity:** No loops or multiple edges among the same pair of vertices.
- **Clear representation:** Easily visualized and understood.

Advantages:

- **Simplicity**: Easier to analyze and understand.
- **Wide applicability**: Suitable for modeling various real-world scenarios.

Disadvantages:

- **Limited representation**: Cannot represent multiple connections between the similar pair of vertices.

2. Multigraph

A **multigraph** allows **multiple edges** between the same pair of vertices.

Example:

A transportation network where multiple routes might exist between two cities, represented by multiple edges between their corresponding vertices.

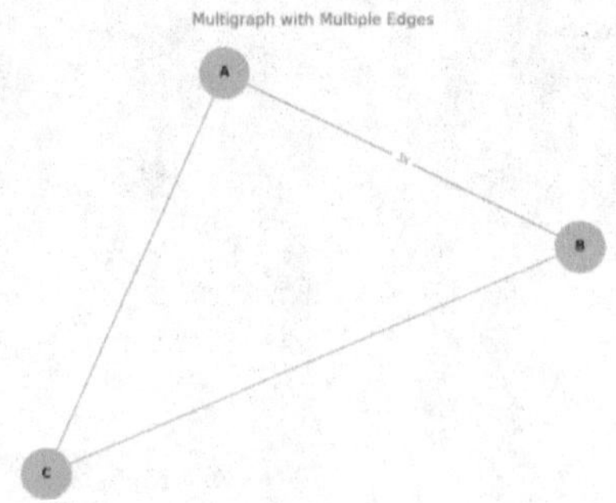

Figure 1.2 | Multigraph Representation with Edge Multiplicity.
Vertices (Nodes):
A, B, C represent individual entities.
Edges (Lines):
- **Multiple Edges:** There are **three edges** between vertices A and B, highlighted as "3x."
- **Single Edges:** A single edge exists between B and C, and another between C and A.

Properties:
- **Multiple edges**: Over one edge can attach the same pair of vertices.
- **Represents parallel connections**: Useful for scenarios where multiple relationships exist.

Advantages:
- **Flexibility**: Allows for more complex relationships between vertices.
- **Real-world representation**: Accurately models scenarios with multiple connections.

Disadvantages:
- **Complexity**: More challenging to analyze and visualize.

3. Directed Graph (Digraph)

A **directed graph** (or digraph) has **directed edges**, meaning the relationships between vertices have a specific direction.

Notation:
- $v_i \rightarrow v_j$: An edge directed from vertex v_i to vertex v_j.

Example:

A one-way street network where edges represent streets, and the direction of the arrow indicates the permitted direction of travel.

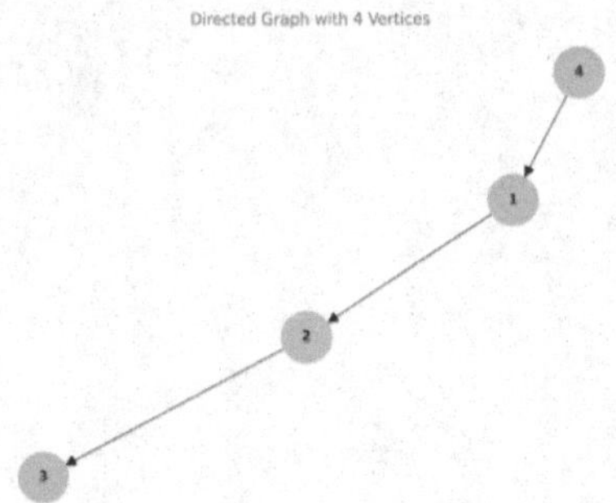

Figure 1.3 | Directed Graph with 4 Vertices

Vertices (Nodes):

1, 2, 3, 4 represent distinct entities.

Edges (Arrows):

- **1 → 2:** Indicates a directed relationship from vertex 1 to vertex 2.
- **2 → 3:** Indicates a directed relationship from vertex 2 to vertex 3.
- **4 → 1:** Indicates a directed relationship from vertex 4 to vertex 1.

Properties:

- **Directionality**: Edges have a precise direction, representing the flow of information or relationship.
- **Asymmetric relationships**: The relationship between vertices can be different in each direction.

Advantages:

- **Real-world representation**: Models scenarios where relationships have directionality.
- **Information flow**: Useful for analyzing information flow and dependencies.

Disadvantages:

- **Complexity**: Requires careful consideration of directionality during analysis.

4. Undirected Graph

An **undirected graph** has **bi-directional edges**, meaning the relationship amongst vertices is symmetrical.

Example:

A friendship network where edges represent friendships, and the relationship is mutual (if A is a friend of B, then B is also a friend of A).

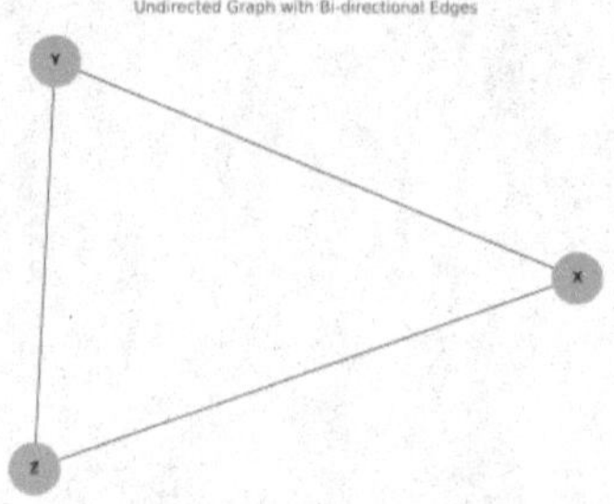

Figure 1.4 | Undirected Graph with Bi-directional Edges.

Vertices (Nodes):

X, Y, Z represent distinct entities.

Edges (Lines):

- **X-Y: Represents a bi-directional relationship between vertices X and Y.**
- **Y-Z: Represents a bi-directional relationship between vertices Y and Z.**
- **Z-X: Represents a bi-directional relationship between vertices Z and X.**

Properties:

- **Bi directionality**: Edges do not have a specific direction.
- **Symmetric relationships**: The relationship between vertices is the same in both directions.

Advantages:

- **Simplicity**: Easier to analyze and visualize.
- **Represents mutual relationships**: Suitable for modeling scenarios where relationships are bidirectional.

Disadvantages:

- **Limited representation**: Cannot represent relationships with directionality.

5. Weighted and Unweighted Graphs

- **Unweighted Graph**: Edges do not have associated values or weights.

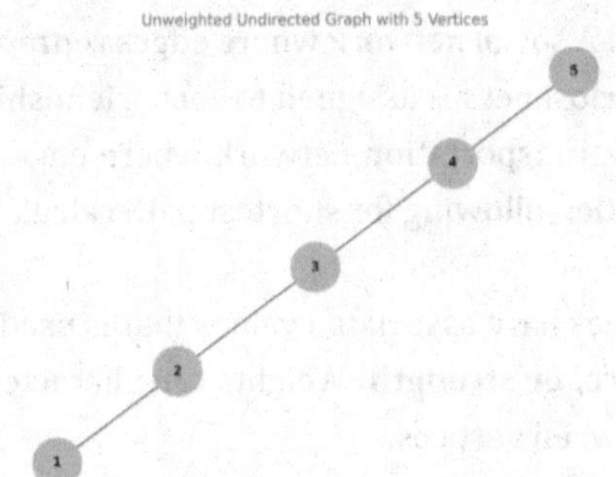

Figure 1.5 | Unweighted Undirected Graph with 5 Vertices (Nodes):
1, 2, 3, 4, 5 represent distinct entities.

Edges (Lines):

- **1-2:** Represents a connection between vertices 1 and 2.
- **2-3:** Represents a connection between vertices 2 and 3.
- **3-4:** Represents a connection between vertices 3 and 4.
- **4-5:** Represents a connection between vertices 4 and 5.

Weighted Graph: Edges have associated weights, representing the cost, distance, or strength of the relationship between vertices.

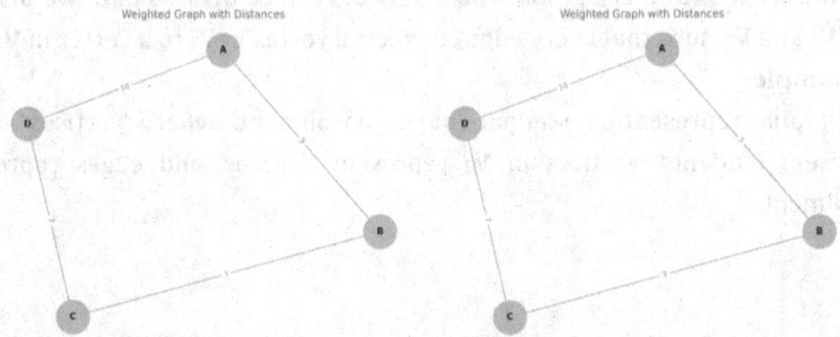

Figure 1.6 | Weighted Graph Representing City Distances

Vertices (Nodes):

- A, B, C, D represent cities.

Edges (Lines):

- **A-B (5):** Distance of 5 units between cities A and B.
- **B-C (3):** Distance of 3 units between cities B and C.
- **C-D (7):** Distance of 7 units between cities C and D.
- **A-D (10):** Distance of 10 units between cities A and D.

Example:

- **Unweighted**: A social network where edges represent friendships, but no specific strength or closeness is assigned to each friendship.
- **Weighted**: A transportation network where edge weights represent the distance between cities, allowing for shortest path calculations.

Properties:

- **Weights**: Edges have associated values that is used for analysis.
- **Cost, distance, or strength**: Weights can characterize various aspects of the relationship between vertices.

Advantages:

- **Real-world representation**: Models scenarios where relationships have associated values.
- **Optimization**: Useful for solving difficulties involving shortest paths, minimum costs, etc.

Disadvantages:

- **Complexity**: Requires additional considerations when analyzing weighted graphs.

6. Special Graphs

a. Bipartite Graph:

A **bipartite graph** is a graph whose vertices can be divided into two disjoint sets, V_1 and V_2, such that every edge connects a vertex in V_1 to a vertex in V_2.

Example:

A graph representing student-course enrollment, where vertices in V_1 represent students, vertices in V_2 represent courses, and edges represent enrollment.

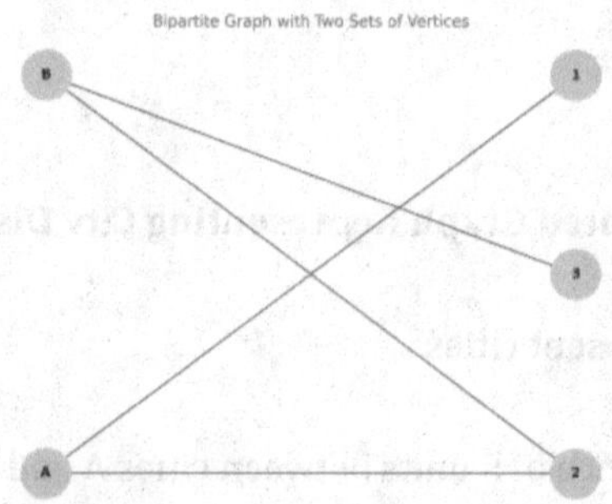

Figure 1.7 | Bipartite Graph with Two Sets of Vertices

Vertices (Nodes):

- **Set V1:** A, B represent entities in the first set.

- **Set V2:** 1, 2, 3 represent entities in the second set.

Edges (Lines):

- **A-1:** Connection between A (V1) and 1 (V2).
- **A-2:** Connection between A (V1) and 2 (V2).
- **B-2:** Connection between B (V1) and 2 (V2).

Properties:

- **Two disjoint sets**: Vertices are divided into two distinct sets.
- **Edges connect sets**: Edges only connect vertices from different sets.

Advantages:

- **Specialized representation**: Suitable for modeling relationships between two distinct sets of entities.
- **Matching problems**: Often used for solving matching problems, where the goal is to find optimal pairings between elements of the two sets.

Disadvantages:

- **Limited applicability**: Not suitable for scenarios where relationships exist in the same set of vertices.

b. Complete Graph:

A **complete graph** is a graph where every pair of vertices is connected by an edge.

Notation:

- K_n: Represents a whole graph with n vertices.

Example:

A social network where everyone is friends with everyone else, with an edge connecting each pair of individuals.

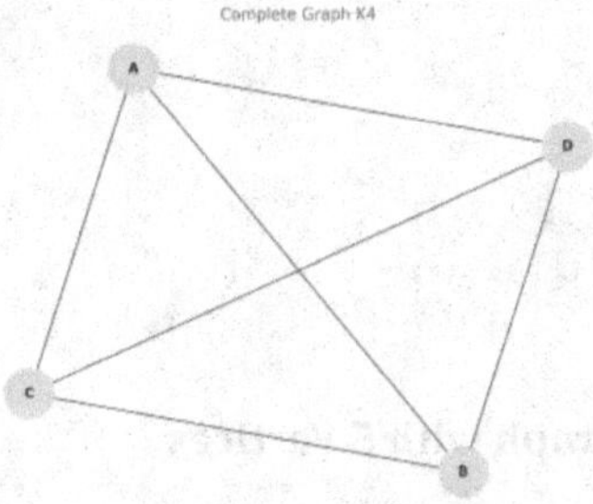

Figure 1.8 | Complete Graph K₄ with 4 Vertices

Vertices (Nodes):

A, B, C, D represent entities.

Edges (Lines):

- Every couple of vertices is connected, resulting in all possible edges:
  - **A-B, A-C, A-D**
  - **B-C, B-D**
  - **C-D**

Properties:

- **All edges**: Every possible edge exists in the graph.
- **Maximum connectivity**: Vertices have the highest possible degree.

Advantages:

- **High connectivity**: Represents scenarios where all entities are directly connected.
- **Mathematical analysis**: Used in various mathematical proofs and algorithms.

Disadvantages:

- **Limited real-world applicability**: Rarely encountered in real-world scenarios.

c. Cyclic Graph:

A **cyclic graph** is a graph containing at least one **cycle**. A cycle is a path that starts and ends for same vertex, with no repeated vertices or edges.

Example:

A road network where a circular route exists, allowing you to start and end for same location while passing through other locations.

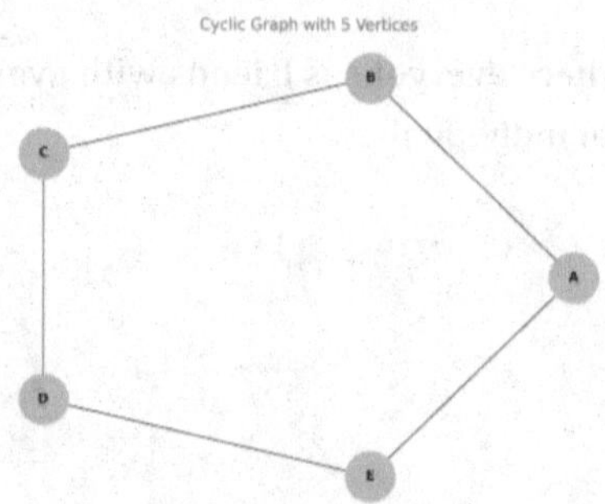

Figure 1.9 | Cyclic Graph with 5 Vertices

Vertices (Nodes):

A, B, C, D, E represent entities.

Edges (Lines):

- The edges form a closed loop, connecting vertices in the sequence:
  - **A-B, B-C, C-D, D-E, E-A**

Properties:
- **Cycles**: Contains at least one closed path.
- **Circular relationships**: Represents scenarios where relationships form a closed loop.

Advantages:
- **Real-world representation**: Models scenarios with circular dependencies or relationships.
- **Pathfinding**: Used in pathfinding algorithms to identify loops and avoid repetitive paths.

Disadvantages:
- **Complexity**: May be difficult to analyze due for presence of cycles.

Examples

Simple Graph:
- **Social Network**: Vertices represent entities, edges represent friendships.

Multigraph:
- **Transportation Network**: Vertices represent cities, edges represent roads, with multiple edges between cities indicating different routes.

Directed Graph:
- **One-way Street Network**: Vertices represent intersections, edges represent streets, with arrows indicating the direction of traffic flow.

Undirected Graph:
- **Friendship Network**: Vertices represent individuals, edges represent friendships, with bidirectional edges indicating mutual friendships.

Weighted Graph:
- **Transportation Network**: Vertices represent cities, edges represent roads, with edge weights representing distances between cities.

Bipartite Graph:
- **Student-Course Enrollment**: Vertices in V_1 represent students, vertices in V_2 represent courses, edges represent enrollment.

Complete Graph:
- **Social Network**: Vertices represent individuals, with every pair connected by an edge, indicating everyone is friends with everyone else.

Cyclic Graph:
- **Road Network**: Vertices represent intersections, edges represent roads, with a circular route forming a cycle.

1.2 GRAPH REPRESENTATIONS (MATRIX AND LIST)

Graphs, as fundamental data structures, are used to model relations and connections between entities. To facilitate efficient handling and analysis of these structures, various representations are employed. Among the most common are the adjacency list and adjacency matrix representations.

Adjacency Matrix Representation

Definition and Structure

An adjacency matrix involves square matrix used to represent a graph. It is an $n \times n$ matrix, where n is - number of vertices in the graph. Each row and column corresponds to a unique vertex. The elements of the matrix, denoted as a_{ij}, represent the presence/absence of an edge between vertices i and j.

Elements

- **Unweighted Graphs:** In unweighted graph, the element a_{ij} is 1 if an edge between vertices i and j, and 0 otherwise.

- **Weighted Graphs:** In a weighted graph, a_{ij} represents the edge weight between vertices i and j if an edge exists, and 0 otherwise. If no edge connects the vertices, the corresponding entry is 0.

Symmetry in Undirected Graphs

For undirected graphs, there is symmetric adjacency matrix. This means that $a_{ij} = a_{ji}$ for all i and j. This symmetry reflects the bidirectional nature of edges in undirected graphs.

Example

Consider a small undirected graph with four vertices, A, B, C, and D, connected by the following edges:

- A-B
- A-C
- B-D
- C-D

The adjacency matrix for this graph is:

	A	B	C	D
A	0	1	1	0
B	1	0	0	1
C	1	0	0	1
D	0	1	1	0

For instance, the entry in row A and column B is 1, indicating thforre an edge between vertices A - B. Similarly, the entry in row C and column D is 1, representing the edge between C and D.

Pros and Cons

- **Pros:**

o **Efficient for Dense Graphs:** In graphs with a high density of edges, adjacency matrices can be computationally efficient for certain processes like checking if an edge exists.

o **Fast Edge Lookup:** Finding the edge between two vertices is straightforward, requiring only a constant time $O(1)$ operation.

- **Cons:**

o **Space Requirements:** An adjacency matrix requires $O(n^2)$ space, making it inefficient for sparse graphs with few limits compared for number of vertices.

o **Memory Consumption:** For large graphs, the memory consumption can be significant.

Adjacency List Representation

Definition and Structure

An adjacency list depiction stores the graph as a collection of lists, one for each vertex. Each list contains the vertices adjacent for corresponding vertex.

Example

For the same graph through vertices A, B, C, and D, the adjacency list representation would be:

- **A:** [B, C]
- **B:** [A, D]
- **C:** [A, D]
- **D:** [B, C]

This representation shows that vertex A is head-to-head to vertices B and C, vertex B is adjacent to A and D, and so on.

Directed vs. Undirected Representation

In a directed graph, the adjacency list for a vertex only includes the vertices that it points to. For example, if there is a focused edge from A to B, then B will be in the list for A, but A will not be in the list for B.

Advantages

- **Space Efficiency:** Adjacency lists are space-efficient for sparse graphs, as ay only store the existing edges.

- **Ease of Neighborhood Traversal:** Traversing the neighborhood of a vertex is easily achieved by iterating through its corresponding list.

Limitations

- **Edge Lookup:** Unlike adjacency matrices, finding an edge among two specific vertices might require iterating through the list of one of the vertices, leading to a time complexity of $O(n)$ in the worst case.

Comparison

Property	Adjacency Matrix	Adjacency List
Space Complexity	$O(n^2)$	$O(n + m)$
Edge Existence Check	$O(1)$	$O(n)$
Suitability for Dense Graphs	Better	Worse
Suitability for Sparse Graphs	Worse	Better

Space Complexity: Adjacency matrices require space relative for square of the number of vertices (n^2), while adjacency lists require proportional space for sum of vertices (n) and edges (m). **Edge Existence Check:** Checking if edge exists is faster in adjacency matrices (constant time $O(1)$) compared to adjacency lists (linear time $O(n)$). **Suitability for Different Graph Densities:** Adjacency matrices are more suitable for dense graphs, while adjacency lists are more efficient for sparse graphs.

Applications

- **Dense Graphs:** Preferred Adjacency matrices for dense graphs where the edges number is close for maximum possible ($n(n$-$1)/2$ for undirected graphs). Frequent edge existence checks and operations like matrix multiplication benefit from this representation.

- **Sparse Graphs:** Adjacency lists are more suitable for sparse graphs where the edges number is significantly less than the maximum possible. Their space efficiency and ease of neighborhood traversal make it ideal for applications like shortest path algorithms and graph traversal algorithms.

In conclusion, the choice between adjacency matrices and adjacency lists depends on the precise characteristics of the graph and the intended operations. Understanding the rewards and limitations of each representation allows for informed selection and efficient graph manipulation.

1.3 SPECIAL GRAPHS (COMPLETE AND CYCLIC)

Complete Graphs

Definition: A complete graph, denoted as K_n, is an undirected graph where every pair of distinct vertices is connected by a unique edge. In other words, each vertex is directly connected to every other vertex in the graph.

Edge Count Formula: The number of edges in a complete graph K_n is given by the formula:

n(n-1)/2

Derivation:

To understand this formula, consider the following:

- Each vertex in a complete graph K_n is connected to *n-1* other vertices.
- If we simply multiply *n* by *(n-1)*, we are double-counting the edges since each edge connects two vertices.
- Therefore, we divide the product *n(n-1)* by 2 to get the correct number of edges.

Properties:

- **High Connectivity:** Complete graphs are highly connected, meaning thforre are many paths between two vertices. This makes them robust to failures, as removing a single vertex or edge does not disconnect the graph.
- **Symmetry:** Complete graphs are highly symmetrical. Every vertex has a same degree, and every edge is equivalent to any other edge in the graph.
- **Vertex Degree:** Each vertex in complete graph K_n have degree of *n-1*. This is because it is connected to every other vertex in the graph.

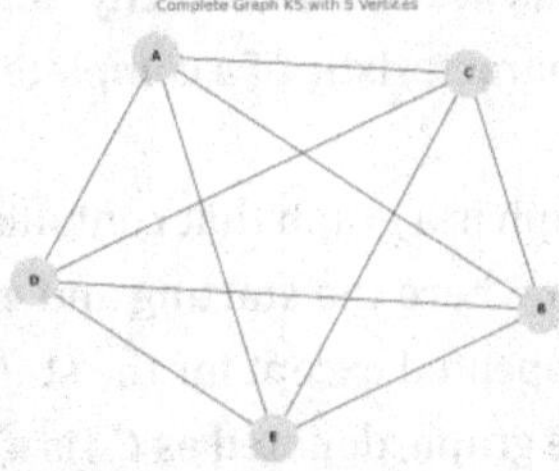

Figure 1.10 | Complete Graph K5 with 5 Vertices

Vertices (Nodes):

A, B, C, D, E represent distinct entities.

Edges (Lines):

- All pairwise connections are present:
  - **A-B, A-C, A-D, A-E**
  - **B-C, B-D, B-E**
  - **C-D, C-E**
  - **D-E**

Applications:

- **Fully Connected Networks:** Complete graphs are model for fully connected networks, where each node can directly communicate with every other node. Examples include peer-to-peer networks and social networks.
- **Clustering Problems:** Complete graphs are used to represent clustering

problems, where the aim is to group similar objects together. Each vertex represents an object, and an edge represents a similarity between two objects. The complete graph allows for exploring all possible connections between objects.

Example:

K₃:

```
  1
 / \
2 3
```

K₄:

```
  1
 / \
2 3
\ /
 4
```

In both examples, every vertex is directly linked to every other vertex, illustrating the defining characteristic of a complete graph.

Cyclic Graphs

Definition: A cyclic graph is a graph that contains at least one cycle. A cycle is a closed path in the Graph where the starting and ending vertices are the same, and no vertex or edge is repeated except for the starting/ending vertex.

Cycle Graph C_n: A cycle graph, denoted as C_n, is a simple cyclic graph having n vertices, where each vertex is associated to two other vertices, forming a single closed loop.

Properties:

- **Symmetry:** Cycle graphs are symmetrical. Each vertex has a same degree (2), and every edge is equivalent to any other edge in the graph.

- **Path Closure:** In a cycle graph, every path can be extended to form a cycle. This is because the graph is closed, and any path can be completed by traversing the remaining edges of the cycle.

- **Acyclic Graphs:** Cyclic graphs are distinct from acyclic graphs, which do not contain any cycles. Acyclic graphs are often called trees or forests.

Types of Cycles:

- **Simple Cycle:** A simple cycle is a closed path within the graph where no vertex or edge is repeated except for the starting/ending vertex.

- **Hamiltonian Cycle:** A Hamiltonian cycle is a cycle that visits every vertex in the Graph every time.

- **Eulerian Cycle:** An Eulerian cycle is that traverses every edge in the Graph for once.

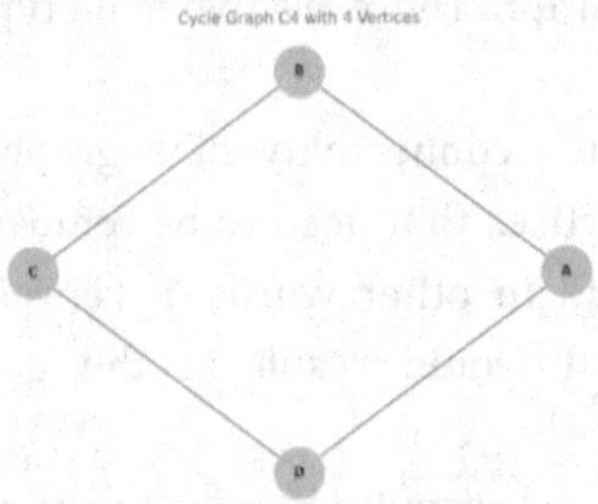

Figure 1.11 | Cycle Graph C4C_4C4 with 4 Vertices

Vertices (Nodes):

A, B, C, D represent distinct entities.

Edges (Lines):

- The edges form a closed loop, connecting vertices sequentially:
- o **A-B, B-C, C-D, D-A**

Applications:

- **Network Routing:** Cyclic graphs belong in network routing algorithms to find efficient paths between nodes. The cycles in the graph represent alternative routes that are used to avoid congestion or failures.
- **Molecular Modeling:** Cyclic graphs are used to model the structure of molecules, where the vertices represent atoms and the edges represent bonds. Cycles in these graphs represent rings in the molecule.
- **Circuit Design:** Cyclic graphs are done in circuit design to represent the flow of current through a circuit. The cycles in the graph represent feedback loops, which can be used to control the behavior of the circuit.

Example:

C_4:

```
1  2
|  |
4  3
```

In this example, each vertex is related to two other vertices, forming a closed loop. This illustrates the structure for cycle graph, where every vertex have degree of 2 and all vertices are part of a single cycle.

1.4 KEY GRAPH PROPERTIES (CONNECTIVITY AND DEGREE)

Connectivity

Connectivity is a important concept in graph theory that measures a graph's robustness and resilience to disruptions. It assesses how well connected the vertices are within the Graph. There are two main types of connectivity:

Vertex Connectivity

Definition: The vertex connectivity of a graph, denoted as $\kappa(G)$, is the minimum number of vertices that need to be removed to disconnect the graph (make it not connected). In other words, it represents the least number of vertices whose removal would result in the graph having at least two components.

Significance: Vertex connectivity quantifies the graph's tolerance to vertex failures. A graph with higher vertex connectivity is more resistant to disruptions caused by the removal of vertices.

Example: For a complete graph K_n, where every vertex is connected to all other vertex. Its vertex connectivity is n-1, as removing any n-1 vertices leaves a single isolated vertex.

Edge Connectivity

Definition: The edge connectivity of a graph, denoted as $\lambda(G)$, is the smallest number of edges that are required to be removed to disconnect the graph. It represents the least number of edges whose removal would result in the graph having at least two components.

Significance: Edge connectivity measures the graph's resilience to edge failures. A graph with higher edge connectivity is more robust to disruptions caused by the removal of edges.

Example: Consider a cycle graph C_n, where vertices are connected in a circular pattern. Its edge connectivity is 2, as removing any two adjacent edges disconnects the graph.

Graph Components

Definition: A connected component of a graph is a maximal subgraph where any two vertices are linked by a path. A graph can have multiple connected components if it is disconnected.

Example: Consider a graph with two disjoint sets of vertices, each forming a complete graph. This graph has two connected components, each representing a complete subgraph.

Examples of Connectivity

Connected Graph:

A simple, fully connected network like a complete graph K$_n$ is an instance of a connected graph. In this graph, every vertex is directly connected to every other vertex.

Disconnected Graph:

Consider a graph with two isolated subgraphs, one being a triangle and the other being a square. This graph is disconnected as are is no path connecting vertices from the triangle for square.

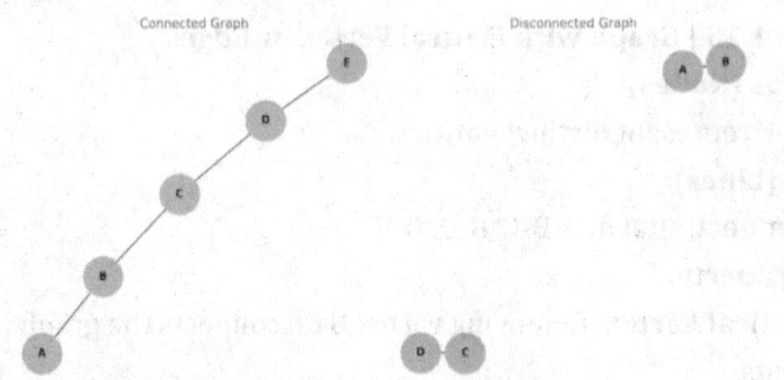

Figure 1.12 | Connected and Disconnected Graphs

Connected Graph:

- **Vertices (Nodes):** A, B, C, D, E
- **Edges (Lines):** A-B, B-C, C-D, D-E
- **Description:** All vertices are connected either directly or via paths, forming a single connected component.

Disconnected Graph:

- **Vertices (Nodes):** A, B, C, D
- **Edges (Lines):** A-B, C-D
- **Description:** The graph consists of two separate components with no edges connecting vertices from different components.

Applications of Connectivity

Network Reliability: Connectivity is vital in network reliability. In communication networks, high vertex connectivity confirms thfor network remains functional after some nodes fail.

Fault Tolerance: In computer systems, connectivity helps design fault-tolerant architectures. By ensuring high connectivity, the system can continue operating even after some components fail.

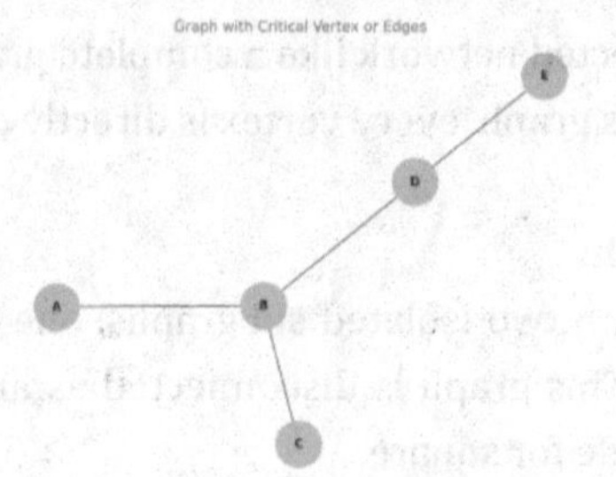

Figure 1.13 | Graph with Critical Vertex or Edges

Vertices (Nodes):

A, B, C, D, E represent distinct entities.

Edges (Lines):

- **Connections:** A-B, B-C, B-D, D-E

Key Property:

- **Critical Vertex:** Removing vertex **B** disconnects the graph into multiple components.

- **Critical Edges:** Removing edges **B-C** and **B-D** also disconnects the graph.

Degree

Degree of a Vertex

Definition: The degree of a vertex in an undirected graph represents the number of edges incident to that vertex. It indicates how many other vertices are directly connected for vertex.

In Directed Graphs: The in-degree of a vertex is the quantity of edges pointing towards it, while the out-degree is the quantity of edges pointing away from it.

Example: In the Graph with 5 vertices, if a vertex has 3 edges connected to it, its degree is 3.

Degree Sequence

Definition: The degree sequence of a graph is a list of the degrees of all its vertices in non-increasing order. It provides a summary of the vertex degrees within the Graph.

Example: For a graph with vertices having degrees 3, 2, 2, 1, 1, its degree sequence is (3, 2, 2, 1, 1).

Andshaking Theorem

Theorem: The sum of the grades of all vertices in graph is equal to twice the number of edges. Mathematically:

$$\sum d(v) = 2|E|$$

where d(v) is the degree of vertex v, and |E| is the number of edges.

Relevance: The Handshaking Theorem highlights a fundamental relationship between vertex degrees and the total number of edges in the Graph. It helps verify the correctness of a graph structure.

Applications of Degree

Central Hubs in Networks: Vertices with high degrees often represent central hubs in networks. In social networks, high-degree vertices correspond to influential individuals with many connections. In biological networks, high-degree proteins may be essential for network function.

Network Analysis: Degree distribution provides insights infor structure and properties of a network. For instance, a power-law degree circulation, where a few vertices have many networks while most have few, is typical of social networks.

Properties Based on Degree and Connectivity

Degree Distributions

Graphs can exhibit various degree distributions, which characterize the spread of vertex degrees within the network.

Power-law Distribution: In real-world networks like social networks and the internet, many vertices have few connections, while a few vertices have number of connections. This distribution follows the power-law, where the probability of a vertex having k connections decays proportionally to $k^{-\gamma}$, where γ is a constant.

Applications and Implications

Social Networks: Degree distributions in social networks reveal the presence of influential individuals (high-degree vertices) and the distribution of connections among users.

Transportation Networks: Degree distributions in transportation networks can indicate the importance of hubs and the flow of traffic.

Network Reliability: The degree distribution can be used to assess the impact of node failures on network reliability. A network with a more uniform degree distribution is generally more resilient to failures.

1.5 FUNDAMENTAL GRAPH OPERATIONS

Union of Graphs

The union of two graphs, denoted as G1 ∪ G2, combines their vertex sets and edge sets. More formally, if G1 = (V1, E1) and G2 = (V2, E2), where V1 and V2 are the vertex sets and E1 and E2 are the edge sets, then the union is defined as V(G1 ∪ G2) = V1 ∪ V2 and E(G1 ∪ G2) = E1 ∪ E2. This operation creates a new graph that retains all vertices and edges from both original graphs.

For example, consider two graphs G1 with vertices {A, B} and edges {AB} and G2 with vertices {B, C} and edges {BC}. The union G1 ∪ G2 will have vertices {A, B, C} and edges {AB, BC}. This union demonstrates how separate components can interact, revealing a broader structure.

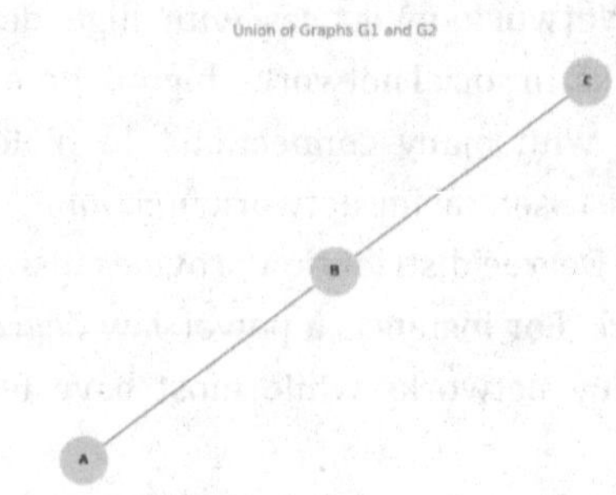

Figure 1.14 | Union of Graphs G1 and G2
Vertices (Nodes):
- G1: A, B
- G2: B, C

Edges (Lines):
- G1: A-B
- G2: B-C

Union Graph Description:
The union graph combines vertices and edges from both G1 and G2, resulting in a single graph where:
- Vertices: A, B, C
- Edges: A-B, B-C

Intersection of Graphs

The intersection of two graphs, denoted as G1 ∩ G2, retains only the shared vertices and edges of the graphs. Formally, the intersection is defined as V(G1 ∩ G2) = V1 ∩ V2 and E(G1 ∩ G2) = E1 ∩ E2. This operation is useful for identifying common features between graphs.

For instance, if G1 has vertices {A, B, C} and edges {AB, BC} and G2 has vertices {B, C, D} and edges {BC, CD}, then G1 ∩ G2 will have vertices {B, C} and edge {BC}. This intersection highlights the common elements that both graphs share, which can be essential in various applications like network reliability.

Complement of a Graph

The complement of a graph G, denoted as G', includes all possible edges not present in the original graph. Formally, if G = (V, E), then the complement is defined by V(G') = V and E(G') = {{u, v} | u, v ∈ V, {u, v} ∉ E }. The complement provides insight infor alternative connections of vertices when certain edges are absent.

In the Graph with vertices {A, B, C} and edges {AB, AC}, its complement would contain edges {BC} since that connection is missing in G. Complement graphs find applications in graph coloring algorithms and clique finding, where understanding the potential connections can guide decisions in complex systems.

Graph Products

Cartesian Product

The Cartesian product for two graphs G1 and G2 is another graph obtained by combining their vertex and edge sets in a specific manner. Formally, if G1 = (V1, E1) and G2 = (V2, E2), the Cartesian product G1 × G2 have vertex set V(G1 × G2) = V1 × V2. An edge exists between vertices (u1, v1) and (u2, v2) if and only if either u1 = u2 and {v1, v2} ∈ E2 or v1 = v2 and {u1, u2} ∈ E1.

This can be visualized in a grid format. For instance, if G1 is a graph with 2 vertices {A, B} and G2 has 2 vertices {1, 2}, G1 × G2 will consist of 4 vertices {(A, 1), (A, 2), (B, 1), (B, 2)} with edges connecting {(A, 1) to (A, 2)} and {(B, 1) to (B, 2)}. Cartesian products are particularly useful in grid-based structures and network design, allowing for complex scenarios to be modeled efficiently.

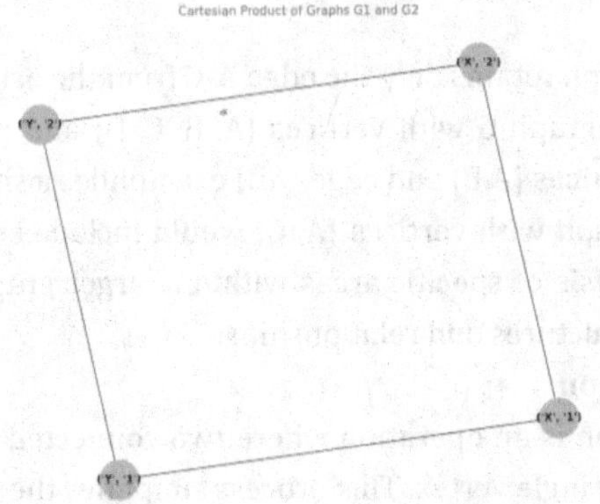

Figure 1.16 | Cartesian Product of Graphs G1 and G2

Vertices (Nodes):

- Cartesian product results in vertices as ordered pairs:

(X,1),(X,2),(Y,1),(Y,2).

Edges (Lines):

- Connections represent the Cartesian product of edges between G1 (X-Y) and G2 (1-2), forming:
  o Edges within (X,1) (X,2), (Y,1), (Y,2).

Subgraphs

Subgraphs are defined, graph formed from a subset of the vertices and edges of another graph. A spanning subgraph includes all vertices of the original graph but only a subset of its edges, while an induced subgraph includes a vertex set and all edges connecting those vertices within the original graph.

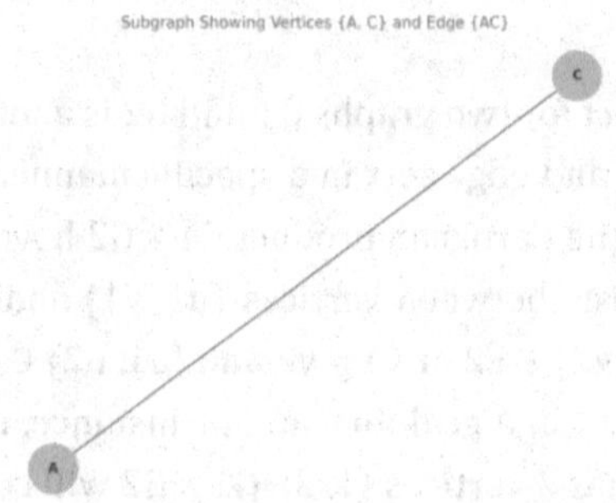

Figure 1.17 | Subgraph with Vertices {A, C} and Edge {AC}

Vertices (Nodes):

- Subgraph includes only vertices **A** and **C**.

Edges (Lines):

- The subgraph retains only the edge **A-C** from the original graph.

For example, a graph G with vertices {A, B, C, D} and edges {AB, AC, CD}. A subgraph with vertices {A,B} and edge {AB} exemplifies a simple subgraph, while an induced subgraph with vertices {A, C} would include edges {AC}. Subgraphs facilitate the analysis of specific areas within a larger graph, enabling targeted examination of structures and relationships.

Edge Contraction

Edge contraction is an operation where two connected vertices in the graph are merged into a single vertex. This process simplifies the graph by reducing its number of vertices and edges while retaining key connectivity properties. Formally, if edge e connects vertices u and v, contracting e replaces u and v with a new vertex w.

This is illustrated by merging vertices in a triangle formed by vertices A, B, and C connected by edges AB, AC, and BC; contracting edge AB results in a new graph with vertices {A/B, C} and edges {A/B-C}. Edge contraction's utility is prominent in optimization problems and simplifies graphs for algorithms, aiding in tasks such as network flow analysis and pathfinding.

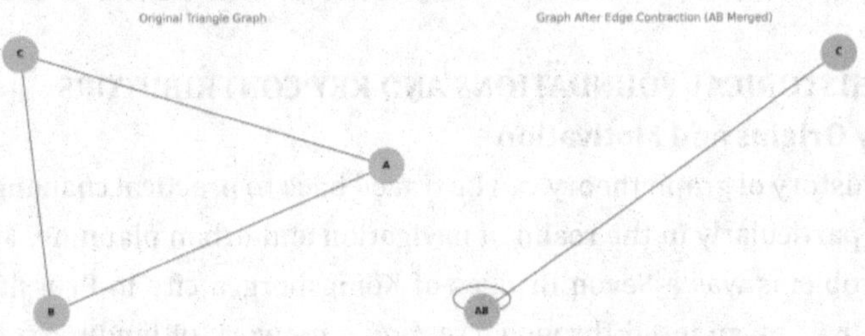

Figure 1.18 | Edge Contraction in a Triangle Graph
Original Graph:
- **Vertices (Nodes):** A, B, C
- **Edges (Lines):** A-B, B-C, A-C

Contracted Graph:
- **Vertices (Nodes):** AB (merged from A and B), C
- **Edges (Lines):** AB-C, maintaining connections from the original vertices.

CHAPTER 2

HISTORY AND APPLICATIONS

2.1 HISTORICAL FOUNDATIONS AND KEY CONTRIBUTORS

Early Origins and Motivation

The history of graph theory can be traced back to practical challenges faced by society, particularly in the realm of navigation and urban planning. Most famous early problems was a Seven Bridges of Königsberg, a city in Prussia where the Pregel River meandered through, creating a network of landmasses and seven bridges connecting them. The challenge posed was to find a walk through the city that would cross each bridge for once without retracing any steps. Mathematician Leonhard Euler took on this problem in 1736, transforming a geographical conundrum into a mathematical one. His profound analysis led for establishment of fundamental concepts in what would become graph theory.

Euler conceptualized the landmasses as vertices (points) and bridges as edges (connections), effectively creating a simple graph representing the city. This is the representation that allowed him to analyze the problem's properties mathematically. Euler concluded that it was impossible to traverse all the bridges under the stated constraints due for properties of vertex degrees-specifically, that more than two vertices with odd degrees would prevent such a path. This foundational work not only solved a local issue but also opened the door for abstract study of networks and connectivity, laying the groundwork for the discipline of graph theory.

Key Mathematicians and Contributions

Following Euler's groundbreaking work, numerous mathematicians contributed for formalization and expansion of graph theory. One notable figure was Gustav Kirchhoff, who in the 19th century applied graph theoretical concepts to electrical engineering. In 1847, he introduced Kirchhoff's circuit laws, which describe how current flows in electrical circuits. By representing circuits as graphs, with vertices as junctions and edges as connections, Kirchhoff's laws became instrumental in analyzing complex electrical networks, further demonstrating the utility of graph theory beyond pure mathematics.

Another pivotal contributor was Arthur Cayley, renowned for his work combinatorial mathematics and algebra. In the 1860s, Cayley introduced trees, a specific type for the graph that plays a crucial role in numerous areas of mathematics, including computer science. His exploration of enumerating distinct trees laid foundational principles for understanding ranked structures and relationships within graphs.

In the 20th century, graph theory experienced significant advancements, owing much for efforts of mathematicians like Oswald Veblen, Philip F. Smith, and Claude Shannon. Veblen's work in topology and other areas bridged gaps in knowledge about graph properties, while Shannon's groundbreaking ideas on information theory and network flow into led for study of the graphs in communication systems, influencing both theoretical and practical aspects of data transmission.

Development Timeline

Graph theory has undergone continuous development since Euler's initial contributions. Below is a timeline highlighting major milestones in the field:

- **1736:** Euler publishes for the solution for Seven Bridges of Königsberg, marking the inception of graph theory.
- **1857:** Kirchhoff formulates laws for electrical circuits, applying graph theory to analyze electrical networks.
- **1877:** Arthur Cayley contributes for understanding of trees, which become vital in combinatorial mathematics.
- **1930:** The concept of directed graphs is introduced, allowing to analyze relationships with more complex interactions.
- **1950s:** The invention of graph algorithms, including Dijkstra's algorithm for finding the shortest path in graphs, expands graph theory's applications in computer science.
- **1960s:** Graph theory applications begin to emerge more prominently in operations research and optimization problems.
- **1980s:** The rise of computational graph theory leads for development of network flows, game theory applications, and advances in quantum computing.
- **21st Century:** Continued evolution in areas such as big data analytics, social network analysis, and biological network modeling, illustrating graph theory's versatility in handling complex systems.

Importance of Graph Theory's Growth

The growth of graph theory has paralleled advancements in mathematics, computer science, and various applied fields. As greater computational capabilities emerged, so did the complexity of systems requiring analysis. The versatility of graph theory made it an crucial tool in operations research, optimization, and computer networking. Graphs provide a natural way for relationships, connections, and interactions among components, thus facilitating problem-solving across diverse domains.

In computer science, the ability to analyze data using graphs has transformed various sectors. Algorithms derived from graph theory are fundamental in routing, data organization, and connectivity analysis. Furthermore, graph theory's connection to machine learning alongwith the artificial intelligence continues to explore novel methods of data representation, revealing patterns and making predictions intrinsic forse technologies.

Examples and Illustrations

Illustrations play a energetic role in comprehending graph theory. The Seven Bridges of Königsberg can be effectively visualized through a diagram where landmasses are labeled as vertices (A, B, C, D) and the edges representing the seven bridges clearly depict the connectivity. Such visualizations not only provide historical context but also introduce critical graph elements, enhancing the understanding of foundational concepts in the subject.

Another classic problem is the Traveling Salesman Problem (TSP), a cornerstone in graph theory and operations research. The TSP seeks the smallest possible route to visits a series of cities (each represented as vertices) and returning origin city. This problem displayed as a weighted graph, where edges represent the distance between cities each pair. The study of TSP has led to various practical applications in logistics and transportation, demonstrating graph theory's continued relevance in solving modern problems.

As technology progresses, contemporary applicable illustrations abound in everyday scenarios. Consider social media networks where users are signified as vertices and their relationships (friendships, follows) as edges. Analyzing these networks utilizing graph theory helps identify influential users, trends, and patterns of interaction, reinforcing the discipline's significance in understanding human behavior and information dissemination.

Real-World Applications of Graph Theory

The graph theory applications extend into numerous fields, showcasing its broad relevancy. In computer networks, graph theory helps design and optimize data flow. By representing networks as graphs, engineers can analyze paths for data packet transmission, assess network reliability, and optimize routing protocols. This application is crucial in the situation of cloud computing and

Internet infrastructure, which rely on robust networking capabilities.

In biology, graph theory assists in modeling complex systems such as protein interactions, which can be characterized as networks to understand cellular processes better. These biological networks help researchers identify potential drug targets, unravel the complexities of diseases, and enhance our understanding of life at a molecular level.

Furthermore, urban planning utilizes graph theory in traffic management and transportation systems. By modeling cities as graphs, planners can analyze and optimize traffic flow, road utilization, and public transport routes, improving overall efficiency and reducing congestion. This application is pivotal for developing smart cities and sustainability initiatives.

2.2 CLASSICAL PROBLEMS IN GRAPH THEORY (SEVEN BRIDGES, FOUR-COLOR THEOREM)

Seven Bridges of Königsberg

The problem of the Seven Bridges of Königsberg is a significant landmark in mathematics and the foundation of graph theory, primarily explored by the mathematician Leonhard Euler in 1736. The city of Königsberg had two islands associated to each other and the continental by seven bridges. The question posed was whether it was possible for walk finished the city crossing each bridge precisely once without repeating any bridge. Euler aimed to solve this problem and, in doing so, established the principles of graph theory.

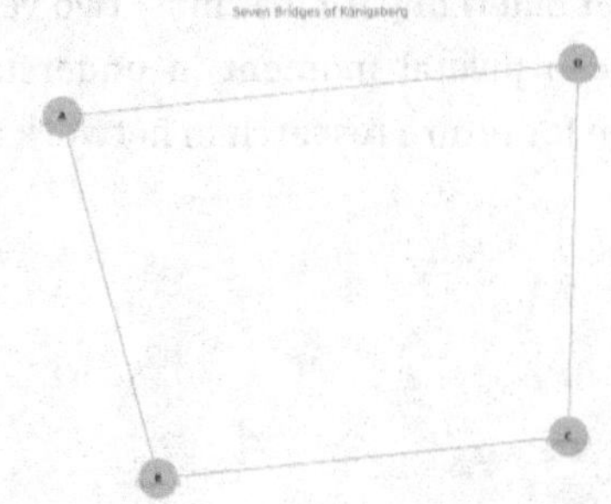

Figure 2.1 | Seven Bridges of Königsberg Representation

Vertices (Nodes):

- A, B, C, D represent the landmasses.

Edges (Lines):

- **A-B:** Two bridges.
- **B-C:** Two bridges.

- **C-D:** Two bridges.
- **A-D:** One bridge.

Euler translated the geographical problem into a mathematical model by on behalf of the landmasses as vertices (points) and the bridges as edges (lines). This abstraction allowed him to analyze the connectivity of the graph. Euler introduced degree of a vertex, which refers for quantity of edges connected to a vertex. He derived an important relation concerning the degrees of vertices in any graph:

$$\Sigma \deg(v) = 2E$$

In this equation:

- **$\Sigma \deg(v)$:** The sum of the degrees of all vertices in the graph.
- **E:** The total number of edges (in this case, the seven bridges).

Using this equation, Euler calculated the degrees of each vertex corresponding for four landmasses (labeled as A, B, C, and D). The observations were:

- Vertex A (3 connections), Vertex B (5 connections), Vertex C (3 connections), Vertex D (3 connections).

Following this, Euler noted that for a path to traverse all edges for once (now called an Eulerian path), there could be at greatest two vertices with an odd degree. In the case of the Seven Bridges, all four vertices had odd degrees, leading for conclusion that such a walk was impossible. Thus, Euler established two critical insights: a graph can have a Eulerian circuit when all vertices have even degrees and can have an Eulerian path if at most two vertices have odd degrees. This revelation marked a pivotal moment in understanding traversability in graphs, setting the stage for future research in network theory and topology.

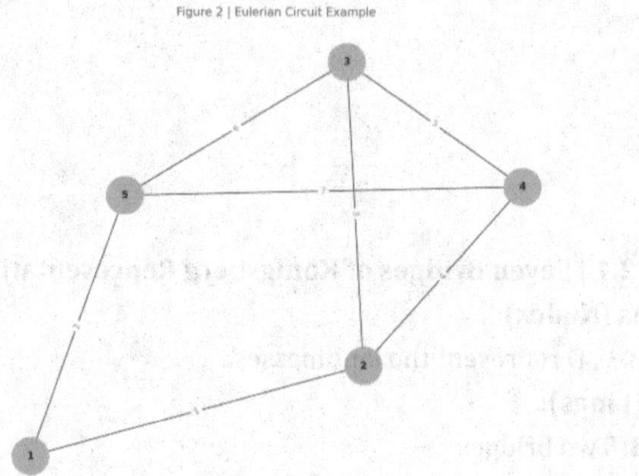

Figure 2 | Eulerian Circuit Example

Figure 2.2 | Eulerian Circuit Example

Vertices (Nodes):

1, 2, 3, 4, 5 represent points of traversal.

Edges (Lines):

- Connections ensure all vertices have even degrees:
  - Edges: 1-2, 2-3, 3-4, 4-5, 5-1, 2-4, 3-5.

Key Property:

- Each vertex have **even degree**, making it likely to traverse all edges for once in a Eulerian circuit.
- Edge labels illustrate the sequence for traversability.

Four-Color Theorem

The Four-Color Theorem presents another classical problem in graph theory, asserting that any planar map can be colored using no more than Four-Color s such that no two adjacent regions share the same color. This theorem illustrates the intersection of combinatorial graph theory and mathematics, demonstrating how theoretical concepts can yield practical applications in map design and geography.

The conjecture first arose in the 1850s through the exploration of how to color maps of counties in England without adjacent areas sharing the same color. Mathematician Francis Guthrie posed the question, leading to various attempted proofs over the decades. However, until 1976 that Kenneth Appel and Wolfgang Haken provided groundbreaking proof using both traditional mathematical arguments and substantial computational verification.

The Four-Color Theorem can be expressed through graph theory concepts, where the map regions represented as graph vertices, with edges denoting adjacency. The main assertion can be formalized as:

$$\chi(G) \leq 4$$

In this equation:

- $\chi(G)$: The chromatic number of a graph G, representing the minimum number of colors needed to color the vertices such that no two adjacent vertices share the same color.

Appel and Haken's proof was notable for its reliance on a computer to check a vast number of configurations and cases, totaling over 1,400 cases, to ensure the theorem's assertion held true for all planar graphs. While some mathematicians were initially critical of the proof's reliance on computational methods, it ultimately demonstrated the theorem's validity and has been widely accepted in the mathematical community.

The implications of the Four-Color Theorem extend beyond mathematics, particularly impacting fields such as cartography, geographic information systems (GIS), and network theory. In practical applications, it aids in optimizing map design, ensuring that adjacent regions are easily distinguishable, thereby enhancing the ease of understanding geographical data.

Furthermore, in scheduling problems, the four-color theorem conceptually applies when assigning resources or time slots to avoid conflicts, similar to how colors are assigned to prevent adjacency clashes. This broad applicability showcases the theorem's historical significance and modern relevance in various domains.

Examples and Influence on Modern Graph Theory

The Seven Bridges of Königsberg and the Four-Color Theorem are fundamental problems that have greatly influenced the development of graph theory. The concepts presented by Euler regarding vertex degrees and paths have led for formulation of algorithms used in various modern applications, including routing and network design. For instance, the principles of traversability guide computer scientists in developing efficient algorithms for traversing networks.

Similarly, the Four-Color Theorem has inspired advancements in graph coloring problems that arise in computer networks, helping to optimize resource allocation and frequency task in wireless communications to prevent signal interference. It exemplifies how a theoretical problem can have profound implications for practical issues in technology and science.

2.3 Applications in Computer Science and Biology

Computer Science Applications

Data Structures

Graph-based data structures form the backbone of many algorithms and data management systems in computer science. Fundamental structures such as trees and adjacency lists allow for efficient representation and manipulation of relationships between data points. Trees, a graph type where one vertex serves as a root, provide hierarchical organization, making operations like searching, insertion, and deletion more efficient. For example, binary search trees allow for $O(\log n)$ time complexity for these operations, showcasing how effective tree structures can optimize data handling.

Adjacency lists offer a memory-efficient way to represent sparse graphs. In an adjacency list, each vertex keeps a list of connected vertices, which provides quick access to neighboring nodes, crucial for many algorithms. This representation is chiefly effective in scenarios where graphs have a significantly lower number of edges compared for quantity of vertices, allowing for compact storage. Consequently, graph data structures facilitate a variety of functions essential to computer applications, including database indexing and memory management.

Algorithms

Graph algorithms play a prominent role in processing data and solving complex computational problems. In search engines, algorithms like PageRank use graph structures to rank web pages based on the links (edges) from other pages. Each page is represented as a vertex, while links between them form the edges. PageRank calculates the importance of a page by considering the quantity and quality of links, emphasizing how graph theory underpins the functionality of modern search engines.

Another significant application pertains to shortest path problems, addressed by algorithms such as Dijkstra's and the A* algorithm. Dijkstra's algorithm efficiently finds the smallest path from the source to all other vertices in the Graph with non-negative edge weights, making it vital for applications in navigation systems, logistics, and route optimization. The A* algorithm enhances this by introducing heuristic methods to expedite the search process further, often used in AI and robotics for pathfinding and movement planning.

In the context of network flows, the Ford-Fulkerson method and its adaptations are critical for solving maximum flow problems in networks, which can represent anything from transportation networks to data flow in computer networks. These algorithms evaluate how resources can be optimally distributed across a network while respecting capacity constraints, often leading to significant improvements in efficiency and performance.

Computer Networks

Graph theory provides essential frameworks for modeling computer networks, focusing on aspects like efficiency and connectivity. In telecommunications, networks are typically characterized as graphs where routers and switches correspond to vertices and the connections between them are edges. This graph-based approach allows network designers to analyze connectivity and bottlenecks, facilitating optimized data transmission paths.

For example, connections in a local area network (LAN) can be imagined as a graph, where the arrangement of endpoints impacts overall network performance. Using algorithms derived from graph theory, network administrators can optimize configurations to improve load balancing, reduce latency, and increase redundancy. Furthermore, concepts like spanning trees are employed to ensure that all nodes in a network are connected with minimal wiring, crucial for cost-effective network infrastructure design.

Biology Applications

Genomics

In biology, particularly in genomics, graphs provide powerful tools to model the complex relationships inherent in genetic data. For instance, sequence alignment techniques utilize graph representations to depict the relationships between DNA or protein sequences, facilitating the documentation of homologous regions. Algorithms such as a Needleman-Wunsch and Smith-Waterman utilize dynamic programming strategies to traverse graphs created from sequence data, emphasizing how graph structures streamline the alignment process.

Gene networks, representing interactions between genes alongwith their products, are also modeled using graphs. Nodes represent genes, while edges signify interactions or regulatory influences. Such graphical representations permit researchers to unveil intricate pathways in cellular processes, helping to identify genetic contributions to diseases and traits. For instance, employing graph-theoretical approaches in studying gene expression data can lead to insights into cancer mechanisms and help for targeted therapies.

Ecology

Graph theory extends its applications into ecology by modeling species interaction networks and food webs. In these graphical representations, species are nodes while interactions, such as predation or competition, form the connecting edges. Analyzing these networks helps ecologists comprehend ecosystem dynamics and biodiversity. For example, the robustness of an ecosystem can be evaluated by examining the connectivity of its species; higher connectivity may indicate greater resilience to environmental changes.

Food webs, representing the flow of energy and nutrients through an ecosystem, can also be analyzed using graph theory. By studying the structure of food webs, researchers can identify keystone species, understand trophic dynamics, and assess the impact of species loss on the ecological community. Advanced metrics derived from graph theory, such as centrality measures, further elucidate the roles different species play within the network, contributing to our understanding of ecosystem health and stability.

Examples and Case Studies

Real-world examples illustrate the profound influence of graph theory in both computer science and biology. In computer networking, the concept of network topology-a description of the organization of various elements (links, nodes, etc.)-is grounded in graph structures. For instance, the Internet itself can be observed as a vast graph where websites are vertices and hyperlinks are edges. Analyzing this topology helps identify critical nodes that, if removed, could disrupt network communication globally, emphasizing the importance of resilience in design.

In the field of biology, one prominent case study involves protein interaction networks, where proteins are modeled as nodes and interactions between them as edges. Understanding these networks is crucial for unraveling the functionality of cellular processes and can ultimately lead to advancements in drug design. By mapping and analyzing these interactions, researchers can identify targets of potential drug and explore pathways involved in diseases such as Alzheimer's and cancer.

2.4 Applications in Social Sciences

Social Networks

Graph theory is a powerful tool for modeling social networks, through complex relationships between individuals and groups. In this context, nodes represent individuals while edges signify the relationships or connections between them. By applying graph-based methods, researchers can elucidate the structure and dynamics of social interactions.

Centrality is a core concept in social network analysis, referring for node importance within the Graph. Various measures of centrality provide different perspectives on an individual's influence or significance:

- **Degree Centrality:** This is the simplest form, calculated as a number of direct connections a node has. It can be expressed as:

Degree Centrality (DC) = k

where **k** is the degree of the node, or the direct edges number connected to it.

Nodes with high degree centrality are often seen as influential, as ay can reach many other nodes directly. For example, in the social network of friends, a person who knows many others with high degree centrality and be well positioned to spread information quickly.

- **Betweenness Centrality:** This measure evaluates the extent to which how each node lies on the shortest paths connecting other nodes. It can be mathematically represented as:

Betweenness Centrality (BC) = $\Sigma\ (\sigma_{st}(v)\ /\ \sigma_{st})$

where $\sigma_{st}(v)$ is the number of shortest paths from node **s** to node **t** that pass through node **v**, and σ_{st} is the total number of shortest paths from **s** to **t**. Nodes with high betweenness centrality may serve as brokers within a network, controlling the flow of information.

- **Closeness Centrality:** This metric calculates how close a node is to all other nodes in the network. It can be quantified as:

Closeness Centrality (CC) = $1\ /\ \Sigma\ (d(v, w))$

where $d(v, w)$ is the shortest distance between nodes **v** and **w**. A node with high closeness centrality can effectively access and disseminate information throughout the network.

Community detection is another crucial aspect of social networks, identifying clusters of closely related nodes. Techniques such as modularity optimization and the Louvain method allow researchers to detect communities within large networks, revealing how social groups form and interact. Understanding communities can assist in various applications, from targeted marketing to political campaigning, where identifying subgroups can optimize outreach strategies.

Moreover, influence in social graphs often stems from the interconnectedness of individuals. The diffusion of information or behaviors through a network can be modeled using these relationships. The concept of "the influence maximization problem" seeks to identify a set of nodes that can effectively spread information for supreme number of other nodes. Algorithms like the greedy algorithm and linear threshold model are usually used to address these challenges, supporting decision-making in marketing and public relations, where understanding the influence of social interactions can drive strategies.

Economics and Political Science

Graph theory is integral to modeling economic and political systems, allowing for the imagining and analysis of complex interactions. For example, trade networks can be represented using graphs where countries or regions are nodes, and the trade relationships between them are the edges. In this context, the analysis can reveal trade patterns, dependencies, and the impact of policy changes on economic interactions.

One of the significant properties often analyzed within such networks is centralization. A centralized network contains one or few nodes (hubs) that dominate the connections. The Degree Centralization (DC) of a graph can be calculated as:

Degree Centralization (DC) = (max DC - DC) / (max DC - min DC)

where **max DC** is the maximum possible degree centrality for any node in the network, and **min DC** is the minimum. High centralization may indicate vulnerabilities; for instance, if the central hub at a trade network is disrupted, it can significantly impact the entire economy.

In political science, graphs can model voting behaviors, allowing for the exploration of how different groups influence election outcomes. Each voter considered a node, and their connections (e.g., friendships or shared political views) as edges. Analyzing this data can reveal voting blocs and how information flows through social circles, affecting campaigns and election strategies.

For example, using graphs to analyze legislative voting records can uncover patterns of coalitions and influence relationships among politicians. By examining relationships through graph-based methods, political analysts can have deeper insights into power dynamics, legislative efficiency, and voter engagement.

Psychology and Sociology

Graph theory also provides insights into interpersonal relationships and group dynamics in psychology and sociology. Social networks are paramount for studying group cohesion, social influence, and the information spread. For instance, group cohesion, often measured by the density of connections within a network, can be analyzed within the Graph framework. A dense network indicates a high level of interaction and communication among members, suggesting stronger cohesion. This can be quantified using the density formula:

Density (D) = 2E / (N(N-1))

where **E** is the quantity of edges and **N** is the nodes number in the graph. A higher density correlates with increased solidarity and support among group members.

Social influence, crucial in understanding how behaviors and norms spread within a community, can likewise be modeled through graphs. The information spread can be analyzed using diffusion models, and the SIR (Susceptible, Infected, Recovered) model is a common epidemiological approach adapted for social contexts. This model illustrates how information spreads through a network, allowing researchers to simulate scenarios of influence and contagion:

$$\lambda = \beta(SI) - \gamma(R)$$

where λ represents the rate of change of the infected individuals, β is the rate of transmission, S is the number of susceptible individuals, I is the number of infected individuals, and γ is the recovery rate.

Understanding the dynamics of information spread enables better strategies in public health messaging, political campaigning, and rumor control. For example, during health crises, using graph theory can help identify key individuals (nodes) who can effectively disseminate public health messages through their connections, maximizing outreach and impact.

Examples and Impact

Several specific examples showcase how graph-based analyses have influenced decision-making, marketing, and policymaking across social sciences. In social media platforms like Facebook and Twitter, user interactions are mapped as graphs, facilitating our understanding of social dynamics and trends. For instance, during social movements, graphs can illustrate how information spreads rapidly, helping activists strategize their outreach efforts.

In marketing, businesses analyze customer social networks to determine key influences within their products' ecosystems. Identifying individuals with high centrality in product-specific networks allows companies to target these influencers for endorsements, significantly impacting consumer behavior and enhancing visibility.

In policymaking, graph theory aids in strategizing responses in public health emergencies. By analyzing contact networks during outbreaks, health officials can predict potential spread paths and implement more effective containment measures. The submission of community detection algorithms can reveal vulnerable populations who may need focused intervention, thus optimizing resource allocation.

2.5 Broader Impact on Various Fields

Multidisciplinary Influence

Graph theory evolved into a foundational framework that intersects with a multitude of disciplines, showcasing its versatility and broad applicability. Fields such as logistics, traffic flow, electrical engineering, and linguistics have harnessed graph-based principles to enhance functionality and optimize processes.

In logistics, graph structures represent transportation networks, where nodes symbolize distribution centers and edges are represented by routes. This visualization allows companies to analyze route efficiency, minimize costs, and optimize supply chains. For example, the Traveling Salesman Problem (TSP), a classic optimization problem, is often tackled using graph algorithms to determine the most efficient route that visits each location for once and returns to its origin.

Traffic flow analysis, similarly, leverages graph theory by modeling roads as edges and intersections as nodes. Traffic management systems utilize algorithms to analyze congestion patterns, optimize traffic signals, and design better road networks. For instance, the use of Dijkstra's algorithm can help in real-time route navigation systems, ensuring the fastest paths are indicated to drivers.

In electrical engineering, circuit design utilizes graph theory to optimize and analyze circuits. Nodes represent components, such as resistors or capacitors, while edges illustrate the connections between them. Techniques derived from graph theory, such as network flow analysis, help engineers streamline circuit layouts and minimize energy losses, ultimately leading for creation of more efficient electronic devices.

Linguistics employs graph theory model language structures, where nodes represent words or phonemes and edges represent relationships or transformations. This approach aids in understanding syntactic structures, analyzing the evolution of languages, and facilitating natural language processing applications. Dependency grammar models, for instance, utilize graph representations to depict the dependencies between words in sentences, enhancing parsing accuracy in linguistics software.

Modern Applications and Technological Advancements

Graph theory's relevance extends into emerging technologies where it plays vital role in artificial intelligence, quantum computing, and cryptography. AI, particularly in natural language processing (NLP) and machine learning, employs graph-based methods to enhance performance. For instance, knowledge graphs represent entities and their relationships within a dataset, allowing AI systems to reason over connected information, thus improving interactions in applications like virtual assistants and chatbots.

In the dominion of quantum computing, graph theory assists in developing quantum algorithms that solve complex problems more efficiently than classical counterparts. Quantum walks, for instance, utilize graph structures to model the behavior of quantum particles, offering new approaches to optimization and search problems in quantum information theory.

In cryptography, graph theory underpins various secure communication protocols. For example, certain cryptographic schemes exploit the hardness of graph-related problems, such as a Graph Isomorphism Problem, to develop secure encryption methods. This integration emphasizes graph theory's significance in safeguarding sensitive information in an increasingly digital world. Data science relies heavily on graph-based methods aimed at network analysis, recommendation systems, and clustering techniques. Social networks, citation networks, and biological networks are analyzed using graph algorithms to extract meaningful insights. Recommendation systems, such as those used by streaming services, utilize collaborative filtering methods that often rely on graph-like structures to suggest content based on user preferences and similar user behaviors.

Clustering algorithms, like community detection methods, categorize nodes in the Graph based on connectivity. This is chiefly useful in identifying groups within social networks, market segmentation, or even biological classifications. Techniques like the Louvain method allow for efficient detection of community structures, facilitating a profounder understanding of network dynamics.

Future Directions

The future of graph theory applications appears promising, with potential expansions in fields such as neuroscience, space exploration, and smart city planning. In neuroscience, graphs can model neural networks, illustrating the connectivity between neurons. Understanding synaptic connections through graph analysis can provide insights into brain functioning and contribute to advances in treating neurological disorders. Space exploration presents unique challenges that can benefit from graph theory's modeling capabilities. Networks representing spacecraft communication, astronaut paths on planetary surfaces, and resource distribution can be optimized using graph algorithms. These methods can ensure efficient operations during missions and inform route planning for future exploration endeavors.

Smart city planning, critical for urban development, will increasingly rely on graph-based models to develop sustainable and efficient infrastructures. Graph theory can help in the design of transportation networks, utility systems, and public services, enabling urban planners to optimize resource allocation and enhance connectivity while minimizing environmental impact. This systematic approach ensures that future urban environments are both livable and efficient.

Impact Examples

The graph theory impact lies across various fields is evidenced through notable examples of efficiency, accuracy, and innovation. In logistics, companies like UPS implement graph algorithms for optimizing delivery routes, resulting in significant fuel savings and improved service times. By analyzing vast delivery networks, they can ensure that each route is streamlined, showcasing the practical graph theory application in real-world operations.

In transportation systems, cities like Los Angeles employ graph-based models to analyze traffic patterns. By understanding congestion points and flow dynamics, city planners can redesign intersections and implement better traffic signal timing, leading to smoother traffic flows and reduced travel times. This application not only enhances commuter experience but also to lower emissions contributes in urban areas.

Graph theory's impact is further evident in social media platforms. Algorithms that

CHAPTER 3

EXPLORING MAGIC LABELING

3.1 Concepts and Development of Magic Labeling

Definition and Basic Concepts

Magic Labeling in graph theory is a type of labeling that assigns values (typically positive integers) for vertices, edges, or both in the Graph such thfor labels follow a particular "magic" condition. This magic condition generally involves achieving a constant sum for the labels associated with each vertex or edge, or across the graph as a complete.

Magic labeling divides into several types, each defined by how labels are applied:

1.   Vertex Magic Labeling: Assigns values to vertices such sum of labels for all vertices head-to-head to an edge equals a fixed constant.

2.   Edge Magic Labeling: Assigns values to edges so that each vertex's total sum of incident edge labels equals a fixed constant.

3.   Total Magic Labeling: Combines vertex and edge labeling is as such that each vertex and its associated edges yield the same magic sum.

The term **"magic"** comes from its similarity to magic squares, where the rows, columns, and diagonals add up to total. This uniformity, when applied to graphs, creates balanced, symmetrical structures that find applications across mathematical modeling, combinatorial designs, and even practical applications.

The foundational idea behind magic labeling is to create numerical harmony within the Graph. This harmony lends itself to modeling balanced structures in networks, making it particularly relevant for fields requiring symmetry, optimization, or error-detection.

Historical Development

The roots of magic labeling are traced back to ancient studies in magic squares, where numbers are arranged in a grid such sums of rows, columns, and diagonals are identical. The magic squares study dates back as early as 200 BCE in China with the Lo Shu Square, and similar concepts were explored in India by mathematicians like Varahamihira.

The leap from magic squares to magic labeling in graph theory started as mathematicians began exploring combinatorial designs and number theory within graph structures. The field of graph theory itself was formalized by Euler in the 18th century of the famous Seven Bridges of Königsberg problem. While Euler didn't study magic labeling directly, his work placed the groundwork for structured approaches in graph theory.

In the 20th century, mathematicians began focusing on applying similar "magic" conditions to graphs. One area of early exploration was Latin squares, which are related to magic squares and combinatorial designs. Studies on these squares led to further interest in how numbers could be arranged across graph structures to achieve balanced sums. The formal concept of magic labeling in graph theory gained traction in the mid-20th century as researchers developed specific labeling techniques. By the 1970s, researchers were examining particular types of magic labeling, including edge magic and vertex magic. This period also saw the development of algebraic and combinatorial techniques to construct magic labeled graphs, allowing mathematicians to extend magic properties beyond simple graphs and apply them to larger, more complex structures.

Over the years, advances in discrete mathematics and combinatorics propelled further research into magic labeling, with new techniques and applications emerging. Today, magic labeling remains an active area of research, with open problems and challenges still being investigated, especially in fields like network theory, error correction, and combinatorial optimization.

Properties and Importance

Magic labeled graphs exhibit several unique properties that distinguish them from other graph types:

1. **Uniform Sum Property:** The defining feature of magic labeled graphs is the constant sum condition, meaning each vertex or edge grouping yields the same sum. This property creates a form of numerical symmetry that is valuable for constructing balanced structures in both theoretical, applied settings.

2. **Structural Constraints:** Not all graphs can be labeled as magic. The structure of the graph, such as its degree or whether it's regular, plays a important role in determining its potential for magic labelability. Regular graphs, for instance, often lend themselves more easily to magic labeling due forir uniformity in vertex degree.

3. **Combinatorial Complexity:** Constructing magic labeled graphs, particularly for complex or large graphs, is often a combinatorial challenge. It involves balancing sums across various configurations, which requires

sophisticated construction techniques, such as those derived from modular arithmetic or graph decomposition.

4. **Applications in Symmetry and Optimization:** Magic labeled graphs are inherently symmetrical, making them valuable in fields that require balanced structures, like network optimization (ensuring even load distribution) and puzzle design (creating balanced and solvable challenges).

The importance of magic labeling extends to both theoretical and real-world applications:

1. In theoretical graph studies, magic labeling helps mathematicians explore properties of symmetry, balance, and unique sum constraints.

2. Magic labeled graphs provide insights into structural graph properties and help solve combinatorial problems.

3. In real-world applications, magic labeling found in network theory, where balanced and symmetrical graphs ensure optimal routing and load balancing.

4. Magic labeling principles are also applied in coding theory to create error-correcting codes, as a constant-sum requirement aids in identifying and correcting errors.

Examples of Magic Labeled Graphs

To illustrate magic labeling, consider the following basic examples:

1. **Simple Path Graph P_3:** A path graph with three vertices, v_1, v_2, and v_3, connected in a straight line by two edges, e_1 and e_2. In vertex magic labeling, each vertex is allocated a label so sum of labels for adjacent vertices around each edge equals a constant. An example labeling for P_3 could assign values such thfor labels of vertices at each edge equal a constant. For instance, setting $v_1 = 1$, $v_2 = 2$, and $v_3 = 1$ achieves a constant sum along each edge.

2. **Complete Graph K_4:** In a complete graph with four vertices, v_1, v_2, v_3, and v_4, magic labeling can be applied by assigning each edge a unique integer to achieve a constant sum at each vertex. If each edge is branded with numbers from 1 to 6 in a balanced configuration, each vertex's incident edges could sum for same constant, satisfying edge magic labeling.

3. **Cycle Graph C_4:** A cycle graph C_4, where four vertices form a closed loop, can achieve magic labeling by assigning values to both vertices and edges. For instance, labeling both edges and vertices so sum around each vertex and along each edge yields a consistent magic total. This makes cycle graphs useful in puzzles and symmetry applications.

Through these examples, we can observe how magic labeling brings a unique balance to graph structures, achieving constant sums across vertices or edges. These properties make magic labeling a fascinating area within graph theory, blending symmetry with combinatorial challenges.

3.2 Types of Magic Labeling (Edge and Vertex)

Vertex Magic Labeling

Definition of Vertex Magic Labeling: In vertex magic labeling, the focus is on assigning values to a graph's vertices such as sum of labels around each edge (or in certain cases, across adjacent vertices) equals a specified constant. In other words, each vertex is labeled in such that for any given edge, the values of the vertices it connects yield the same total sum. This constant sum condition is what makes the labeling "magic" and creates a unique balance within the graph.

To define this more formally, consider a graph $G = (V, E)$ where: V is the set of vertices. and E is the set of edges connecting these vertices.

If each vertex $v \in V$ is assigned a label from a set of integers, then vertex magic labeling requires that for any edge $e = (v_i, v_j) \in E$, the sum of the labels of v_i and v_j equals a constant k across the graph. This labeling often imposes significant constraints on the graph's structure and on the labeling itself.

Conditions and Limitations: Certain conditions determine whether a chart can achieve vertex magic labeling:

1. **Graph Regularity:** Vertex magic labeling is often easier to achieve in regular graphs, where each vertex has a same degree. This regularity simplifies the balancing act of ensuring a constant sum for all edges.

2. **Graph Type:** Vertex magic labeling tends to work well in specific graph types, such as cycles or paths, where the structure naturally supports balanced vertex sums.

3. **Label Constraints:** The labels themselves may need to fall within a specific range or follow certain properties (like being consecutive integers), as ase constraints can directly impact the feasibility of obtaining a consistent sum across edges.

Due forse structural and numerical requirements, vertex magic labeling is limited in scope and may not be possible for all graphs, especially irregular or highly complex graphs.

Edge Magic Labeling

Definition of Edge Magic Labeling: Edge magic labeling, in contrast to vertex magic labeling, focuses on the edges rather than vertices. In edge magic labeling, each edge is assigned a label so sum of all edge labels incident for the vertex equals a fixed constant across the graph. This constant vertex sum ensures that each vertex have balanced sum of adjacent edge labels, creating the magic effect.

For a graph $G = (V, E)$ with vertices V and edges E, edge magic labeling assigns values to each edge $e \in E$ such that, for each vertex $v \in V$, the sum of the labels of edges incident to v equals a constant k.

Differences from Vertex Magic Labeling: Edge magic labeling differs from vertex magic labeling in the following key aspects:

1. **Labeling Focus:** Edge magic labeling assigns values to edges, whereas vertex magic labeling assigns values to vertices.

2. **Graph Constraints:** Edge magic labeling may apply to a wider range of graphs, particularly those where the vertex degree is not uniform, as a edge sums are independent of vertex labeling.

3. **Feasibility:** Edge magic labeling can be more flexible than vertex magic labeling since edge sums are usually easier to balance across vertices. However, the graph's degree distribution and structure still play a role in determining whether it can be edge magic.

Comparison Between Edge and Vertex Magic Labeling

To better understand the distinctions between edge magic and vertex magic labeling, here's a side-by-side comparison highlighting their unique characteristics:

Feature	Vertex Magic Labeling	Edge Magic Labeling
Definition	Vertices are labeled so that adjacent vertices' sums are constant.	Edges are labeled such that incident edge sums at each vertex are constant.
Labeling Focus	Vertices	Edges
Graph Requirements	Often requires regular graphs or specific structures.	Applicable to a wider variety of graphs, not necessarily regular.
Common Graph Types	Cycles, paths, and certain regular graphs.	More versatile, suitable for trees, multigraphs, and general graphs.
Applications	Useful in symmetry-based applications and combinatorial designs.	Often applied in network optimization and balanced load distribution.

Examples of Magic Labeling

Example 1: Vertex Magic Labeling in a Path Graph P_3:

Consider a simple path graph P_3 with three vertices v_1, v_2, and v_3 connected by two edges e_1 and e_2:

Assign labels $v_1 = 1$, $v_2 = 2$, and $v_3 = 1$.

With this configuration, the sum of vertex labels around each edge e_1 and e_2 equals a constant, 3. This setup illustrates vertex magic labeling, where each edge maintains the required sum.

Figure 1 | Path Graph P₃ with Vertex Magic Labeling

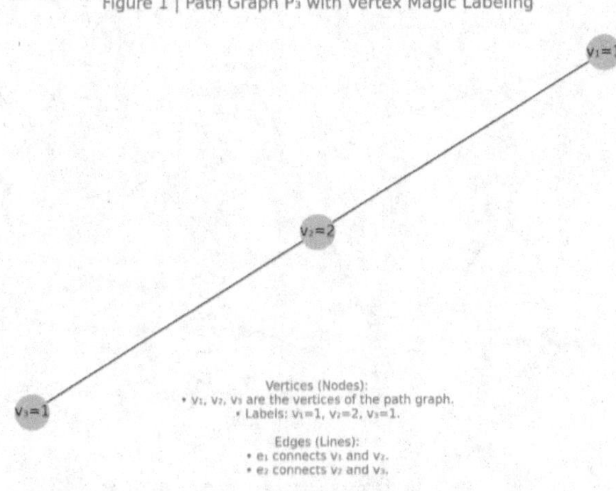

Figure 3.1 | Path Graph P_3 with Vertex Magic Labeling

Vertices (Nodes):

- v1,v2,v3 are the vertices of the path graph.
- Labels: v1=1, v2=2, v3=1.

Edges (Lines):

- e1 connects v1 and v2.
- e2 connects v2 and v3.

Magic Property:

- The sum of labels around each edge equals 3.

Example 2: Edge Magic Labeling in a Complete Graph K_3:

In a complete graph K_3 with three vertices v_1, v_2, and v_3, and edges e_1, e_2, and e_3 connecting each pair of vertices:

Assign each edge a unique label, such as $e_1 = 1$, $e_2 = 2$, and $e_3 = 3$.

With this edge magic labeling, the sum of labels incident to each vertex equals a constant, 3. This demonstrates how edge magic labeling can create balanced vertex sums.

Through these examples, we see how magic labeling, whether applied to vertices or edges, achieves numerical harmony and provides a unique structure within the graph.

Figure 2 | Complete Graph K₄ with Edge Magic Labeling

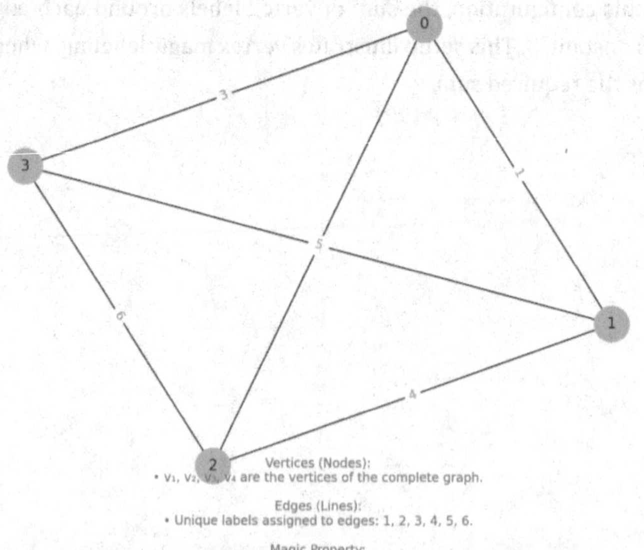

Figure 3.2 | Complete Graph K4K_4K4 with Edge Magic Labeling

Vertices (Nodes):

- v1,v2,v3,v4 are the vertices of the complete graph.

Edges (Lines):

- Unique labels assigned to edges: 1, 2, 3, 4, 5, 6.

Magic Property:

- The sum of labels for edges incident to any vertex equals the same constant.

3.3 Construction Techniques (Algebraic and Combinatorial)

Algebraic Techniques

1. **Modular Arithmetic:** Modular arithmetic plays a central role in constructing magic labeled graphs, particularly when ensuring that sums or products yield consistent results. By working within a modular system (such as modulo n), values wrap around after reaching a certain point, providing a repeating cycle that is useful for maintaining uniformity across a graph. For example, labeling vertices or edges within a modulus can help enforce a constant-

sum requirement, as numbers repeat predictably within a set range. For instance, if we're working within modulo 5, any integer value will cycle back to fit within 0-4. This can ensure sum of labels around vertices or edges remains consistent, supporting the magic properties.

2. **Systems of Equations:** Systems of linear equations are another algebraic tool used in magic labeling. By setting up equations for the relationships between labels, mathematicians can solve values that satisfy magic sum conditions across vertices or edges. For example, if any graph has specific constraints on edge sums, equations can be written to represent these constraints and solved simultaneously. In practice, systems of equations allow for a methodical approach to assigning values that ensure each vertex or edge's sum meets the magic criterion. For example, graph with three vertices v_1, v_2, v_3 and required sums at each edge, equations can model these relationships to find suitable integer solutions.

3. **Transformations:** Transformations are used to manipulate and adjust values as a way that preserves the magic properties. For instance, arithmetic transformations (like addition or multiplication by a constant factor) can be applied uniformly across vertices or edges to maintain balance. Transformations also make it possible to adapt magic labeled graphs to different scales, allowing larger or smaller graphs to exhibit the same properties by adjusting values systematically.

4. **Algebraic Constructions Based on Latin Squares and Arithmetic Sequences:**
Latin squares-grids in which each row and column contains a unique set of symbols-are closely related to magic labeling. By assigning unique values to a Latin square and interpreting these values as vertex or edge labels, a consistent pattern can be achieved across the graph. This uniform arrangement is particularly useful in generating magic graphs where every row and column (analogous to edges and vertices) yields a balanced sum.

Arithmetic sequences are also beneficial in constructing magic labeled graphs. By arranging values in an arithmetic progression (e.g., consecutive integers), the sums around each edge or vertex can be made consistent. This method used in simpler graphs, like cycles or complete graphs, where the structure naturally supports arithmetic sequences for achieving balanced labels.

Combinatorial Techniques

1. **Recursive Methods:** Recursive methods allow for building magic labeled

graphs incrementally. Starting with a smaller graph that satisfies the magic conditions, vertices or edges can be added recursively while preserving the constant sum requirement. This method is particularly effective for larger graphs, as it allows for a controlled, step-by-step expansion of the graph while maintaining balance. In practice, a recursive method might start with a small cycle or path graph, adding vertices or edges while adjusting labels to retain the magic properties. This gradual approach is especially useful in dynamic structures, where the graph may grow over time.

2. **Graph Decomposition:** Graph decomposition involves breaking down a complex graph into simpler subgraphs, can be labeled independently. By focusing on smaller components, it's easier to assign values that meet the magic criteria, and the labeled subgraphs can then be combined to form a magic labeled graph. For instance, a complex graph may be decomposed into cycles or paths, each labeled separately. Once labeled, these subgraphs are reassembled infor original graph. This method is effective for constructing magic labeled graphs with high degree of complexity or irregular structure, as decomposition simplifies the task.

3. **Combinatorial Design Techniques:** Combinatorial designs, such as balanced incomplete block designs (BIBDs) and other arrangements, are structured frameworks that support consistent labeling across a graph. Combinatorial designs ensure that each label or set of labels appears in a balanced manner, satisfying magic requirements without requiring manual adjustment for each edge or vertex. By applying combinatorial designs, magic labeling can be systematically achieved. For example, BIBDs are used in scenarios where each vertex or edge sum needs to remain balanced across different subsets of the graph. This design structure helps in maintaining consistency and ensures thfor labeling fits within a specific configuration, especially in complex or multi-part graphs.

Step-by-Step Example Construction

Example 1: Algebraic Construction using Modular Arithmetic:

Consider a simple cycle graph C_4 with four vertices v_1, v_2, v_3, v_4.

To create a vertex magic labeling:

Assign Labels in a Modular Sequence: Start by assigning labels 1, 2, 3, and 4 to each vertex, working in modulo 5. This wraps values around, ensuring they repeat within the set range.

Verify Consistent Edge Sums: Check each pair of adjacent vertices to see if the sums match the magic constant. Adjust as necessary to achieve balance while staying within the modulus.

Result: The result is a cycle graph with a constant sum for each edge, achieved through modular repetition.

Figure 3 | Cycle Graph C₄ with Total Magic Labeling

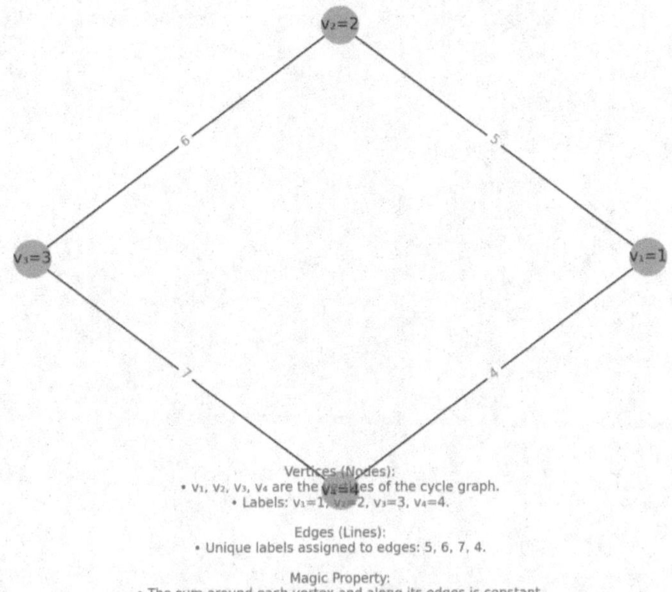

Figure 3.3 | Cycle Graph C4C_4C4 with Total Magic Labeling

Vertices (Nodes):

- v1,v2,v3,v4 are the vertices of the cycle graph.
- Labels: v1=1, v2=2, v3=3, v4=4.

Edges (Lines):

- Unique labels assigned to edges: 5, 6, 7, 4.

Magic Property:

- The sum around each vertex and along its edges is constant.

Example 2: Combinatorial Construction with Recursive Expansion

Consider constructing a magic labeled path graph P_4 with vertices v_1, v_2, v_3, v_4:

Start with P_2: Begin with a path graph P_2 with vertices v_1 and v_2, assigning labels that meet the required sum for the single edge.

Expand to P_3: Add a new vertex v_3, adjusting the labels to ensure the sum around each edge remains consistent.

Continue Expanding: Repefor process to add v_4, adjusting labels as needed.

Result: By expanding recursively and adjusting sums, a balanced, vertex magic-labeled path graph P_4 is achieved.

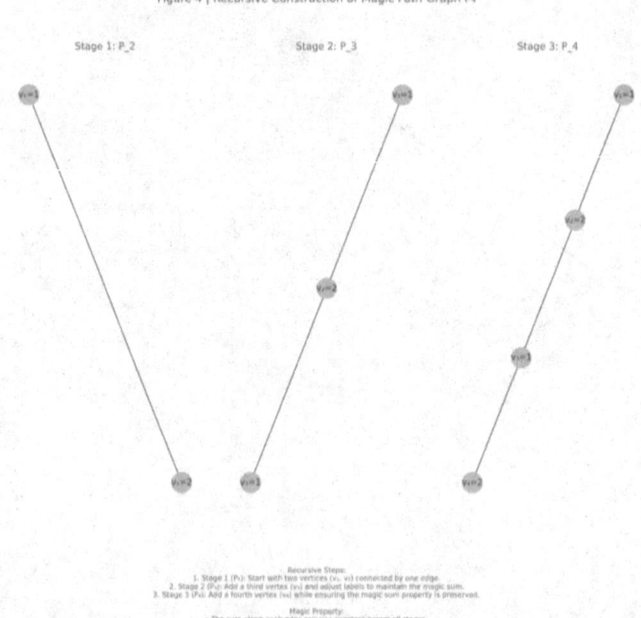

Figure 3.4 | Recursive Construction of Magic Path Graph P4P_4P4

Recursive Steps:

1. **Stage 1 (P_2):** Start with two vertices (v1,v2) connected by one edge.

2. **Stage 2 (P_3):** Add a third vertex (v3) and adjust labels to maintain the magic sum.

3. **Stage 3 (P_4):** Add a fourth vertex (v4) while ensuring the magic sum property is preserved.

Magic Property:

- The sum along each edge remains constant across all stages.

Applications of Construction Techniques

Construction techniques for magic labeled graphs have several real-world applications, particularly in fields that require balanced structures:

1. **Scheduling:** Magic labeling is used to create balanced schedules where tasks or resources need to be evenly distributed, such as in sports tournaments or job rotations.

2. **Optimization:** Network optimization uses magic labeling to ensure that resources, data flow, or load is balanced, reducing bottlenecks and improving efficiency in communication networks.

3. **Design and Patterning:** In design fields, such as circuit design or architectural planning, magic labeling can help in arranging elements symmetrically or creating uniform patterns. Magic labeled graphs ensure balanced, repeatable patterns that meet specific constraints for functionality or aesthetics.

These algebraic and combinatorial techniques provide structured ways to construct magic labeled graphs, meeting the needs of both theoretical studies and practical applications.

3.4 Applications in Puzzles and Coding Theory

Magic labeling, very powerful tool with a extensive range of applications, particularly in recreational mathematics puzzles and coding theory. The principles of balanced labeling and symmetry make magic labeling especially valuable for constructing puzzles and enhancing error detection and correction in digital communication systems.

Magic Labeling in Puzzles

1. **Magic Squares and Recreational Mathematics:** Magic labeling has roots in the ancient concept of magic squares, where numbers are arranged in a square grid where sum of (each) row, column, and diagonal is identical. This arrangement represents an early form of magic labeling, where each cell in the grid can be supposed of as a vertex, and the relationships (sums) between these cells yield a consistent result. Magic squares have fascinated mathematicians for centuries due forir symmetry and balance, serving as a foundation for more complex magic labeling concepts.

2. **Sudoku and Other Numerical Puzzles:** Magic labeling also appears in puzzles like Sudoku, where players must fill a grid such that each row, column, and subgrid contains unique numbers within a specific range. While Sudoku is not typically viewed as a magic labeling problem, the underlying concept of balance and non-repetition in rows and columns aligns with magic labeling principles. Another example is **Latin squares**, a type of combinatorial design where each row and column contains a unique set of values, similar to a magic square but without the diagonal requirement. Latin squares are used in creating puzzles with balanced structures that challenge players to discover patterns and solutions through logic.

3. Other Puzzles: Some puzzles, especially those involving graph structures, use magic labeling explicitly. For example:

Magic Graphs: Graph-based puzzles challenge players to allocate values to vertices or edges such that all adjacent sums are constant.

Magic Star Puzzles: In these puzzles, players assign numbers to points on a star or geometric shape such that each line segment or connection yields the same total. This form of magic labeling is engaging for its symmetry and balance, presenting a satisfying problem in recreational mathematics.

These types of puzzles demonstrate how magic labeling principles add depth to problem-solving, requiring players to balance values and find patterns across the grid or graph structure.

Coding Theory Applications

1. Error-Correcting Codes: Magic labeling is highly applicable in coding theory, especially in constructing **error-correcting codes**. In digital communication, data transmission errors are inevitable due to noise, interference, or hardware faults. Magic labeling provides a way to organize data so that errors can be detected and corrected effectively.

2. Magic-labeled structures ensure balanced arrangements in codewords, where each code segment maintains a consistent sum or balance. If an error occurs (e.g., a bit flip in binary code), the inconsistency in the sum can signal a problem, allowing for the error to be located and corrected.

3. Balanced Codes for Reliable Transmission: Magic labeling also contributes to **balanced code construction**. In a balanced code, each segment of data follows a specific pattern or sum, which can prevent error propagation. This structure is crucial for maintaining data integrity, as balanced codes are less likely to experience cumulative errors.

4. For example, in **Hamming codes** or **Reed-Solomon codes**, magic labeling principles help organize code elements so that data verification and correction are efficient. Magic labeling ensures that transmitted data maintains consistency across segments, improving reliability in data transmission.

5. Applications in Digital Communications: In digital communication, magic labeling is particularly valuable for reducing signal interference and ensuring that transmitted data matches the original source. Balanced labeling prevents imbalances that could distort the signal, a critical factor in both wireless communication and wired data transmission.

Examples and Case Studies

Example 1: Magic Square in Puzzle Design:

Consider a **3x3 magic square puzzle** where the objective is to arrange numbers 1 to 9 so that each row, column, and diagonal sums to 15.

The simplicity of this puzzle demonstrates the appeal of magic labeling: by finding the correct configuration, players create a balanced, symmetrical layout.

This foundational example of magic labeling has inspired countless variations, including larger magic squares and non-standard configurations, all rooted in the same principles of balance and symmetry.

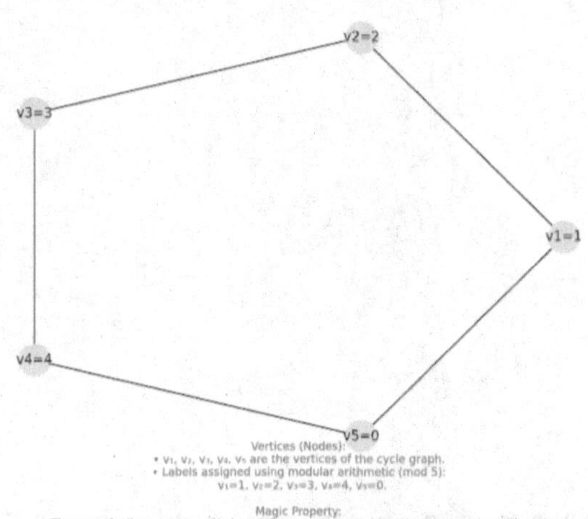

Figure 3.5 | Cycle Graph C₅ with Modular Arithmetic Labeling

Vertices (Nodes):

- v1,v2,v3,v4,v5 are the vertices of the cycle graph.
- Labels assigned using modular arithmetic (mod 5):
- v1=1, v2=2, v3=3, v4=4, v5=0.

Magic Property:

- The sum of adjacent vertex labels satisfies a constant, achieved using modular arithmetic.

Example 2:

Latin Square in Error Correction:

A **Latin square** used in error-correcting codes, where each row and column of a grid contains unique values. If applied to coding theory, a Latin square ensures that each row and column of a transmitted code segment follows a predictable pattern.

This consistency enables error detection, as any deviation from the pattern suggests an error.

For example, values in a specific row or column do not match the expected unique values, the deviation indicates an error. This arrangement provides an efficient mechanism for identifying and correcting errors, which is essential in systems like barcode scanners and data entry verification.

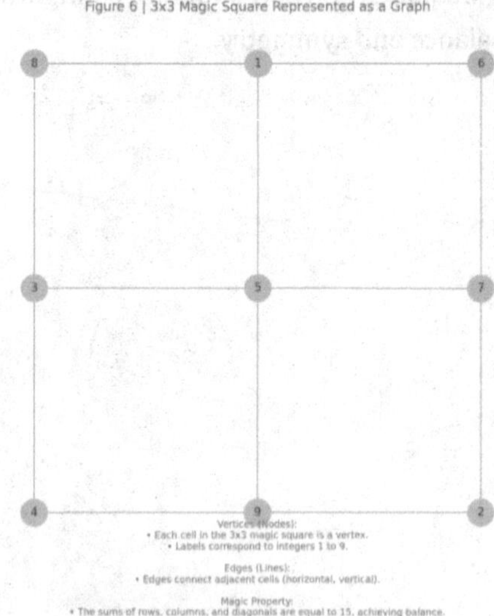

Figure 3.6 | 3x3 Magic Square Represented as a Graph

Vertices (Nodes):

- Each cell in the 3x3 magic square is a vertex.
- Labels correspond to integers 1 to 9.

Edges (Lines):

- Edges connect adjacent cells (horizontal, vertical).

Magic Property:

- The sums of rows, columns, and diagonals are equal to 15, achieving balance.

Example 3: Hamming Code and Magic Labeling in Digital Communication:

In Hamming codes, each bit in the transmitted codeword is carefully labeled and arranged so thfor overall structure supports **parity checking**-a form of error detection that aligns with magic labeling's consistency requirements.

By ensuring that every bit follows a designated pattern, Hamming codes can detect and correct single-bit errors, improving data integrity during transmission.

This balanced approach aligns with magic labeling principles, where symmetry and predictable sums enhance reliability.

Through these examples, we see how magic labeling serves as a opening concept in both puzzles and coding theory. The principles of balance, symmetry, and constant sums are not only mathematically satisfying but also functionally essential in error correction, data verification, and puzzle design. These applications highlight the versatility of magic labeling, demonstrating its relevance from recreational mathematics to complex digital communication systems.

3.5 Challenges and Open Problems in Magic Labeling

Magic labeling is a captivating and intricate area of study in graph theory, offering various theoretical and practical challenges. Although many types of magic labeled graphs have been explored, significant open problems and questions remain. These challenges range from proving the feasibility of magic labeling in complex graphs to classifying different magic-labeled graph types. Additionally, ongoing research aims to expand the application of magic labeling concepts to new fields.

Current Challenges in Magic Labeling

1. Complexity in Large or Complex Graphs: One of the main challenges in magic labeling is constructing magic labels for larger or more complex graphs. As a size and complexity of a graph increase, so does the difficulty of finding a set of values that meet the constant sum requirements for edges or vertices. For instance:

- **Regular vs. Irregular Graphs**: Regular graphs, where all vertices with same degree, are generally easier to label, as air uniform structure simplifies balancing. Irregular graphs, OTOH require more intricate label arrangements.

- **Multigraphs and Hypergraphs**: Multigraphs (graphs with multiple edges between vertices) and hypergraphs (graphs where edges can connect more than two vertices) further complicate the process, as ay require maintaining constant sums across more diverse edge configurations.

2. Proving Existence and Uniqueness: Another critical challenge is proving the existence or uniqueness of magic labelings for certain graph classes. Key questions include:

- **Existence**: For some graphs, it is unknown whether a valid magic labeling can exist at all. This uncertainty complicates research, as determining labelability often requires extensive exploration or new theoretical frameworks.

- **Uniqueness**: Even when magic labeling is possible, valid labelings (unique configurations) may vary. Identifying whether a magic labeling is unique or if multiple configurations exist for the same graph is a challenging task that involves deep mathematical analysis and often demands combinatorial proofs.

3. Computational Limitations: The process of labeling a graph can become computationally intensive, for large graphs or graphs with irregular structures. Since magic labeling often relies on iterative or recursive techniques, computational efficiency is essential. Developing algorithms which handle complex graphs efficiently without exhaustive searches remains an ongoing challenge, particularly when verifying or generating multiple potential labelings.

Unsolved Problems

1. **Classification of Magic Graphs**: A fundamental unsolved problem in magic labeling is the comprehensive classification of all magic-labeled graph types. While certain types, like cycle graphs and paths, have been studied in detail, many graph classes remain unexplored in terms of their magic labelability. Establishing a complete classification system for magic-labeled graphs would provide a foundation for further research and could uncover new insights into graph properties.

2. **Conjectures on Magic Labelability**: Several conjectures in graph theory relate to magic labeling, yet they remain unproven. Some of these conjectures involve: **Magic Labelability of Specific Graph Families**: Certain families of graphs, like bipartite or highly connected graphs, may have specific labelability properties that are yet to be proven or disproven. **Magic Constants and Their Limits**: Understanding the constraints on magic constants (the sums required for magic labeling) in various graph classes could reveal broader principles that govern labelability.

3. **Limitations of Magic Labeling**: The limitations of magic labeling in specific graph classes, such as sparse or dense graphs, are not fully understood. While dense graphs are naturally support magic labeling due forir interconnected nature, sparse graphs often lack sufficient edges to meet the sum requirements. Identifying the structural limitations of magic labeling for different graph types remains a major open problem.

Potential Research Directions

1. **Developing New Construction Techniques**: To address the challenges

in magic labeling, researchers are exploring new construction techniques that could simplify the labeling process. For instance: **Hybrid Techniques**: Combining algebraic and combinatorial approaches may yield more flexible and efficient methods for constructing magic-labeled graphs. **Probabilistic Methods**: Applying probabilistic techniques could allow researchers to explore large numbers of potential labelings in an efficient manner, which could help identify previously unknown magic labelings or support theoretical proofs.

2.   **Extending Magic Labeling Concepts to New Domains**: Magic labeling principles are being explored in fields outside traditional graph theory. For example: **Network Theory**: Magic labeling can be applied to network optimization and load balancing, where balanced sums can represent efficient data distribution across nodes. **Quantum Computing**: In quantum computing, graph structures play a role in designing stable qubit connections, and magic labeling concepts may support error correction and stable state maintenance in quantum circuits.

3.   **Investigating Magic Labeling in Hypergraphs**: Hypergraphs, which allow edges to connect multiple vertices, represent a promising area for future research. Extending magic labeling techniques to hypergraphs could lead to new applications and insights, as hypergraphs model complex systems that standard graphs cannot capture. Research area could support advancements in fields like biology, chemistry, and computer science, where hypergraph models are prevalent.

Examples of Research Impact

1.   **Network Theory and Load Balancing**: Recent research has shown that magic labeling principles can enhance load balancing in network theory. By assigning balanced labels to network nodes, data or task distribution can be optimized to prevent bottlenecks, improving the overall efficiency of network systems. This application demonstrates how a theoretical concept from graph theory can have a tangible impact on technology and resource management.

2.   **Computational Complexity and Algorithm Design**: Advances in magic labeling have contributed to computational complexity theory by influencing algorithm design. Solving magic labeling problems has inspired new algorithms that can handle complex combinatorial tasks, benefiting fields that require large-scale data processing or pattern recognition. For instance, magic labeling concepts have led to innovative solutions in algorithms for network analysis and scheduling.

3. **Error Correction in Digital Communications**: Magic labeling principles have also impacted error correction strategies in digital communications. By utilizing balanced label arrangements, researchers have developed more efficient methods for detecting and correcting transmission errors. This research supports more reliable data transmission, particularly in applications like satellite communication and high-speed internet, where data integrity is crucial.

Through these examples, we see how solving challenges in magic labeling has led to breakthroughs in both theory and practice. Magic labeling not only deepens our understanding of graph theory but also provides valuable tools for addressing real-world problems in fields as varied as computational complexity, communication, and network optimization.

CHAPTER 4

GRACEFUL LABELING

4.1 Introduction to Graceful Labeling and Examples

Definition and Basic Concepts

Graceful Labeling Definition:

Graceful labeling is labeling in graph theory that assigns distinct integer values for vertices of a graph in such way that creates a unique mapping between vertex and edge labels. In a gracefully labeled graph, the goal is to label each vertex with a unique integer from a set of non-negative integers, such that when each edge is labeled with the absolute difference among the labels of its two endpoints, the resulting edge labels are all unique and span the sequence of integers from 1 up for number of edges in this graph.

Formally, consider a graph $G = (V, E)$ with n vertices and m edges. A labeling function f is said to be graceful if it maps each vertex $v \in V$ to a unique integer in the set $\{0, 1, ..., m\}$ such that for each edge $e = (v_i, v_j) \in E$, the edge label defined by $|f(v_i) - f(v_j)|$ results in a unique integer from 1 to m. This ensures that each edge in the graph have unique label and thfor edge labels form a contiguous set of integers.

Importance of Graceful Labeling:

Graceful labeling is particularly significant in combinatorial graph theory due to its applications in network theory, coding theory, and communication systems. The concept was introduced as a way for structure graphs for optimal routing, symmetry, and balance, making it an essential tool for designing network structures. In network topology, graceful graphs enable efficient channel assignment and reduce interference, making them valuable for frequency allocation and error detection in digital communications. Additionally, graceful labeling's systematic nature aids in the construction of structured graphs in mathematical research, enhancing graph analysis and algorithm design.

Properties and Constraints

Unique Properties of Graceful Labeling:

1. **One-to-One Mapping of Vertex Labels to Edge Labels:** A defining property of graceful labeling is the unique mapping between vertex labels and edge labels. Since each vertex label contributes to a distinct edge label through

its difference with adjacent vertex labels, this mapping creates a structured, balanced labeling.

2. **Uniqueness of Edge Labels:** In a gracefully labeled graph, each edge must have a unique label that corresponds for absolute difference between the tags of its endpoints. This requirement means that each edge label, from 1 up to m, must be represented for once. This stuff ensures that no two edges for graph share the same label, which is crucial for maintaining a harmonious and symmetrical structure within the graph.

3. **Contiguous Edge Labels:** Another characteristic of graceful labeling is thfor edge labels are contiguous, meaning thfory cover a sequence of integers without gaps. This trait allows graceful labeling to form a well-ordered pattern that is both aesthetically and mathematically balanced, enhancing the symmetry of the graph.

4.2 Techniques for Graceful Labeling:

Not all graphs can be gracefully labeled. Various graph structures inherently support or restrict graceful labeling:

1. **Paths and Cycles:** Path graphs (linear arrangements of vertices and edges) are known to be gracefully labelable. Similarly, cycle graphs can often be labeled gracefully, although their closed-loop structure imposes additional challenges in achieving unique edge labels.

2. **Trees:** Trees, especially binary trees, are another class of graphs frequently studied for graceful labeling. While some tree configurations support graceful labeling, others, particularly those with irregular or highly unbalanced structures, may not achieve a graceful configuration. A conjecture in graph theory, known to Graceful Tree Conjecture, posits that trees are gracefully labelable, although this conjecture remains unproven for all tree types.

3. **Graph Constraints:** Certain structural constraints, like the degree of vertices and graph connectivity, can affect the feasibility of graceful labeling. For instance, highly dense graphs or those with complex connectivity patterns may not easily meet the requirements for graceful labeling, as a unique and contiguous labeling conditions become difficult to satisfy.

Basic Examples

Example 1: Graceful Labeling of a Path Graph P_4:

Consider a path graph P_4 with four vertices v_1, v_2, v_3, v_4 and three edges connecting them in a straight line. The labeling process can proceed as follows:

1. **Vertex Labeling:** Assign the vertices the labels $f(v_1) = 0$, $f(v_2) = 3$, $f(v_3) =$

1, and $f(v_4) = 4$.

2. **Edge Labeling:** For each edge, calculate the absolute difference between labels of endpoints:

- For edge $e_1 = (v_1, v_2)$, the label is $|f(v_1) - f(v_2)| = |0 - 3| = 3$.

- For edge $e_2 = (v_2, v_3)$, the label is $|f(v_2) - f(v_3)| = |3 - 1| = 2$.

- For edge $e_3 = (v_3, v_4)$, the label is $|f(v_3) - f(v_4)| = |1 - 4| = 3$.

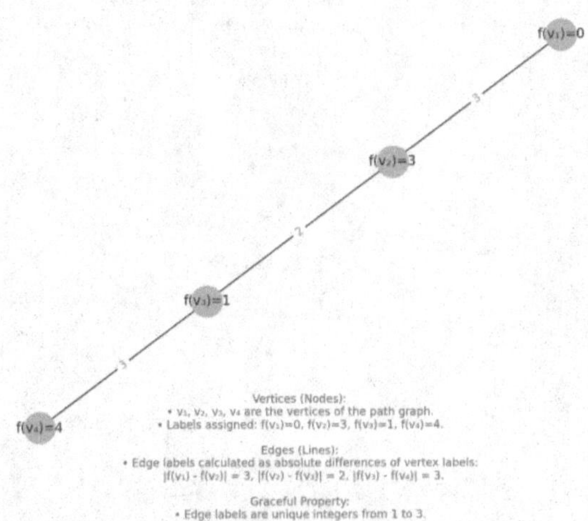

Figure 1 | Path Graph P₄ with Graceful Labeling

Figure 4.1 | Path Graph P4P_4P4 with Graceful Labeling

Vertices (Nodes):

- v1,v2,v3,v4 are vertices of the path graph.

- Labels assigned: f(v1)=0, f(v2)=3, f(v3)=1, f(v4)=4.

Edges (Lines):

- Edge labels calculated as absolute differences of vertex labels:

o |f(v1)−f(v2)|=3

o |f(v2)−f(v3)|=2,

o |f(v3)−f(v4)|=3.

Graceful Property:

- Edge labels are unique integers spanning the set {1,2,3}.

This configuration satisfies graceful labeling since each edge have unique label in the sequence {1, 2, 3}, with no two edges sharing the same label.

Example 2: Graceful Labeling of a Simple Cycle Graph C_4: Consider a simple cycle graph C_4, with vertices v_1, v_2, v_3, v_4 arranged in the closed loop and edges e_1, e_2, e_3, e_4 connecting each consecutive pair of vertices:

1. **Vertex Labeling:** Assign the vertices labels $f(v_1) = 0$, $f(v_2) = 2$, $f(v_3) = 4$, and $f(v_4) = 1$.

2. **Edge Labeling:** Calculate the edge labels:

   - $e_1 = (v_1, v_2)$: $|f(v_1) - f(v_2)| = |0 - 2| = 2$.

   - $e_2 = (v_2, v_3)$: $|f(v_2) - f(v_3)| = |2 - 4| = 2$.

   - $e_3 = (v_3, v_4)$: $|f(v_3) - f(v_4)| = |4 - 1| = 3$.

   - $e_4 = (v_4, v_1)$: $|f(v_4) - f(v_1)| = |1 - 0| = 1$.

Figure 2 | Cycle Graph C₄ with Graceful Labeling

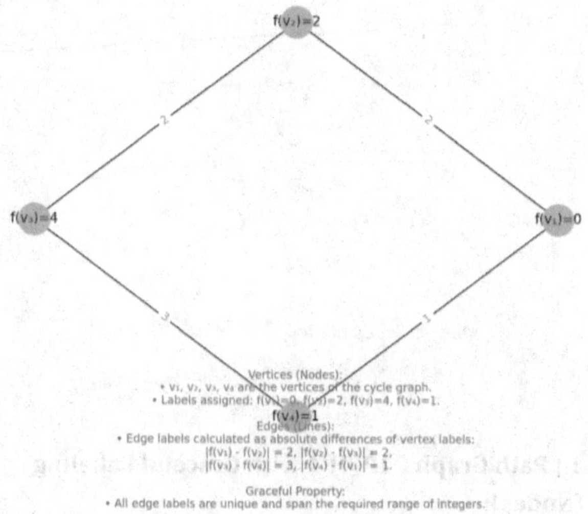

Figure 4.2 | Cycle Graph C4C_4C4 with Graceful Labeling

Vertices (Nodes):

- v1,v2,v3,v4 are the vertices of the cycle graph.
- Labels assigned: f(v1)=0, f(v2)=2, f(v3)=4, f(v4)=1.

Edges (Lines):

- Edge labels calculated as absolute differences of vertex labels:
- $|f(v1)-f(v2)|=2$,
- $|f(v2)-f(v3)|=2$,
- $|f(v3)-f(v4)|=3$,
- $|f(v4)-f(v1)|=1$.

Summary of Examples:

These examples prove the systematic approach required for graceful labeling, where each vertex label is carefully chosen to ensure all edge labels remain unique and contiguous. Through examples like path and cycle graphs, we see the fundamental principles of graceful labeling: distinct and non-repeating edge labels, a one-to-one mapping, and a contiguous sequence of edge labels.

Graceful labeling provides a structured framework for balancing vertex and edge labels, supporting requests in network design, coding theory, and other fields requiring consistent, balanced structures. As shown in these examples, graceful labeling achieves symmetry and order within the Graph, making it a valuable tool in mutually theoretical and applied graph studies.

4.3 Applications in Communication Networks

Graceful labeling offers powerful tools for optimizing communication networks, improving data transmission, minimizing interference, and ensuring efficient use of resources. The structured and balanced nature of graceful labeling aligns closely with the needs of modern communication networks, making it a valuable strategy in network design, load balancing, and frequency management.

Network Design and Optimization

Optimizing Signal Paths and Load Balancing: In communication networks, efficient data transmission and load balancing are essential to avoid congestion and reduce latency. Graceful labeling aids in achieving these goals by providing a structured framework for routing and organizing network traffic. Through graceful labeling, vertices (network nodes) and edges (connections or links) are assigned values that create a balanced system, where each link handles an even distribution of traffic, minimizing bottlenecks and preventing overloading in any one area.

1. **Signal Path Optimization**: In a network, signal paths often involve the movement of data packets across multiple nodes from a source for destination. Graceful labeling supports optimal path selection by ensuring that each connection maintains a consistent label difference, which simplifies routing decisions and balances the data flow. As data packets move through the network, this balance helps maintain stable connections and prevents individual paths from becoming overwhelmed. For instance, in a mesh network where each node linked to multiple others, graceful labeling is used to assign values that distribute data packets evenly across all possible paths. This approach prevents specific routes from becoming overburdened, resulting in improved network stability and reduced latency.

2. **Load Balancing**: Load balancing is a critical requirement for communication networks, particularly in data centers and high-traffic networks.

By applying graceful labeling principles, network designers can ensure that data loads are evenly distributed across all nodes and paths. In a gracefully labeled network, the consistent distribution of edge labels enables a balanced load on each link, reducing the risk of network congestion. This is especially important for networks that manage a big number of simultaneous connections, such as cloud computing infrastructures or data centers that serve millions of users. Graceful labeling helps prevent any single link from becoming a bottleneck by routing data along balanced paths, ultimately enhancing the overall efficiency of the network.

3. **Enhanced Routing and Path Selection**: In gracefully labeled networks, routing protocols can income advantage of the balanced structure to make efficient path selections. Since each path in a gracefully labeled graph maintains a predictable label pattern, routers and switches can easily calculate the most efficient route without needing extensive computational resources. This streamlined routing process results in faster packet forwarding and a reduction in processing overhead at each node.

Improving Network Performance Through Balanced Routes: Graceful labeling also contributes to enhanced network performance by facilitating balanced routes for data transmission. Balanced routes help maintain uniform signal strength and quality across the network, which is particularly helpful in large-scale networks or wireless systems where signal degradation can occur over long distances.

1. **Network Stability**: A balanced network where all paths share a consistent level of traffic is inherently more stable than a network with uneven load distribution. In cases where certain paths carry significantly more data than others, network performance may degrade due to overload. Graceful labeling mitigates this risk by providing a consistent framework for load distribution, ensuring that each connection handles a proportional share for overall data load.

2. **Minimizing Packet Loss and Reducing Latency**: Balanced routes also help minimize packet loss by reducing the probability of congestion and signal interference. When graceful labeling is applied for network, data packets are less likely to experience delays or drops, as a load is evenly spread across available connections. This even distribution helps maintain low latency, which is essential for applications requiring real-time data, such as video streaming, online gaming, or telemedicine.

Channel Assignment and Frequency Allocation

Role of Graceful Labeling in Frequency Allocation: In wireless communication networks, channel assignment and frequency allocation are essential tasks that determine the quality and reliability of communication. Graceful labeling assists in this process by reducing interference between channels and enabling efficient frequency distribution across the network.

1. **Minimizing Interference**: In wireless networks, interference occurs when neighboring channels operate on similar frequencies, leading to signal degradation and reduced data quality. By using graceful labeling to assign frequencies, network designers can create a separation between frequencies on adjacent links, minimizing overlap and interference. This approach ensures that each channel operates within a distinct frequency range, preserving signal integrity and improving overall network performance.

2. **Efficient Spectrum Utilization**: Graceful labeling also supports efficient spectrum utilization by allowing network designers to assign frequencies in the way that maximizes the available bandwidth. Since each link have unique label (corresponding to a unique frequency in the spectrum), the network avoids the inefficiencies of overlapping frequencies. This efficient use of spectrum is particularly valuable in environments with limited bandwidth, such as densely populated urban areas or locations with a high concentration of wireless devices.

3. **Example in Cellular Networks**: In cellular networks, graceful labeling principles are used to allocate frequencies to cell towers and mobile devices. Each cell tower covers a specific geographic area (cell), and neighboring cells must operate on non-interfering frequencies to prevent signal overlap. By applying graceful labeling, network planners can assign frequencies that minimize interference across cells, enabling clear, uninterrupted communication for mobile users.

Practical Applications in Channel Allocation for Wi-Fi and Satellite Communication: Graceful labeling is also applied in Wi-Fi networks, where multiple devices connect for same access point. In Wi-Fi environments, interference between devices can cause data loss and slow network speeds. Graceful labeling helps assign channels to devices in the way that minimizes overlap, ensuring a smooth flow of data.

1. **Wi-Fi Networks**: In a Wi-Fi network with multiple access points, graceful labeling can guide the allocation of channels to minimize interference between adjacent access points. By strategically transfer channels based on graceful labeling, network administrators may optimize coverage areas and reduce signal

overlap, resulting in improved network quality and higher data transfer speeds for users.

2. **Satellite Communication**: In satellite communication, graceful labeling principles assist in assigning frequencies to satellites to avoid cross-interference. By ensuring that each satellite operates on a unique frequency, graceful labeling reduces the chance of signal interference and enhances the reliability of long-distance communication.

Examples and Case Studies

1. **Example 1: Data Center Network Using Graceful Labeling for Load Balancing**:

In a large data center, thousands of servers must communicate seamlessly to manage vast amounts of data. To avoid congestion and maintain high performance, network engineers implement graceful labeling to balance the load across all server connections. By labeling each connection uniquely and ensuring an even distribution of data flow, graceful labeling enhances the data center's efficiency, enabling fast, reliable access for users.

2. **Example 2: Cellular Network Frequency Allocation**:

In a densely populated city, a mobile network provider uses graceful labeling techniques to assign frequencies across its cellular towers. Each tower operates on a distinct frequency, reducing the likelihood of signal interference between neighboring cells. This method ensures that users experience uninterrupted service, even in areas with high call and data traffic volumes. By preventing frequency overlap, the provider can support more users within a limited spectrum.

3. **Case Study: Graceful Labeling in Satellite Networks**:

A satellite communication company applies graceful labeling to optimize frequency allocation across its fleet of satellites. Each satellite is assigned a unique frequency, which helps prevent cross-interference as ay relay signals to Earth. By utilizing graceful labeling, the company achieves a clear and stable communication channel for satellite transmissions, improving the reliability of data received on the ground.

4. **Case Study: Graceful Labeling in High-Density Wi-Fi Environments**:

In a crowded convention center, network administrators deploy a Wi-Fi network with numerous access points to serve thousands of attendees. To reduce signal interference and maintain a high-quality connection, they apply graceful labeling to assign channels to each access point. This strategic channel allocation prevents overlap, allowing attendees to connect without experiencing drops or delays. By enhancing network quality, graceful labeling ensures that attendees can access online resources without interruptions.

4.4 Graceful Labeling in Special Graphs

In graph theory, special types of graphs present unique challenges and opportunities for graceful labeling. These graphs, which include paths, trees, bipartite graphs, and more complex structures like multipartite graphs and hypercubes, often have distinct characteristics that influence the feasibility of graceful labeling. This section explores how these special graphs are labeled, the specific techniques used, and the adaptations required to meet the graceful labeling conditions.

Special Graph Types

1. Paths and Cycles:

Path and cycle graphs some of the most straightforward types for graceful labeling.

Path Graphs: A path graph P_n, with n vertices arranged in a straight line, is a commonly studied example of graceful labeling. Path graphs are gracefully labelable due forir simple, linear structure, which makes it easier to assign distinct labels that result in unique edge differences. For instance, in a path graph with four vertices (P_4), the labeling 0, 3, 1, 4 on consecutive vertices creates edge labels $|0 - 3| = 3$, $|3 - 1| = 2$, and $|1 - 4| = 3$, each have unique and meets graceful labelling.

Cycle Graphs: Cycle graphs C_n, which form a closed loop, also support graceful labeling under certain conditions. The structure of cycle graphs adds complexity due for circular arrangement, where the labeling must account for the closure of the graph. This requirement often limits the types of cycle graphs that can be gracefully **labeled, as only cycles of certain sizes can meet the conditions.**

2. Trees:

Trees are a significant area of interest in graceful labeling due forir widespread use in modeling hierarchical data structures, networks, and decision processes.

Binary Trees: A binary tree is a tree structure in which node havet most two children. These trees can often be gracefully labeled by assigning numbers to vertices in the way that ensures each edge's label is unique. For example, labeling the root as 0 and incrementally labeling each level can achieve the necessary balance. The Graceful Tree Conjecture hypothesizes that all trees are gracefully labelable.

Star Graphs: A star graph is a type of tree where one central vertex connects directly to all other vertices, with no further connections among them. Star graphs can be gracefully labeled by assigning 0 for central vertex and sequential numbers for outer vertices. This setup allows for unique edge differences between the center and each outer vertex, meeting the requirements for graceful labeling. Star graphs are one of the simpler tree structures for achieving graceful labeling.

3. Bipartite Graphs:

Bipartite graphs, which consist of two disjoint vertices sets with edges connecting vertices from one set for other, offer a mixed level of suitability for graceful labeling. These graphs can often be gracefully labeled, particularly when structured as paths or trees.

Complete Bipartite Graphs: In complete bipartite graphs, every vertex in one subset is connected with vertex in the other subset. This structure allows for graceful labeling in some cases.

Special Considerations for Bipartite Graphs: In cases where a bipartite graph cannot be gracefully labeled directly, it may be possible to use partial labeling or adapted techniques that approximate graceful properties. These adaptations allow for balanced labeling while retaining as much of the graceful structure as possible.

Graph-Specific Challenges and Adaptations

1. Multipartite Graphs:

Multipartite graphs extend the concept of bipartite graphs by dividing vertices into more than two disjoint sets.

Challenge: Due for complex interconnections across multiple vertex sets, multipartite graphs present substantial challenges for graceful labeling. Ensuring that each edge label remains unique and covers the range of required values becomes increasingly difficult as a subsets number grows.

Adapted Techniques: One approach to gracefully labeling multipartite graphs is to focus on balanced labeling within each subset, rather than attempting to label the entire graph gracefully.

2. Hypercubes:

Hypercubes, also known as n-dimensional cubes, are complex graphs where vertices represent binary strings of length n, and edges connect vertices that differ by exactly one bit.

Challenge: Hypercubes have highly symmetrical and dense structures, making it difficult to apply traditional graceful labeling. The dense interconnectivity of vertices complicates the labeling process, as achieving unique edge labels while covering all **required values is challenging.**

Adapted Techniques:

Detailed Examples

Example 1: Graceful Labeling of a Star Graph S_5:

A star graph S_5 consists of one central vertex v_0 and five outer vertices v_1, v_2, v_3, v_4, and v_5, with each outer vertex connected for center.

Vertex Labeling: Assign the central vertex v_0 a label of 0. Label the outer vertices sequentially with values 1, 2, 3, 4, and 5.

Edge Labeling: Each edge's label is the absolute difference between the endpoints labels:

$$- e_1 = (v_0, v_1) \text{ is labeled } |0 - 1| = 1,$$
$$- e_2 = (v_0, v_2) \text{ is labeled } |0 - 2| = 2,$$
$$- e_3 = (v_0, v_3) \text{ is labeled } |0 - 3| = 3,$$
$$- e_4 = (v_0, v_4) \text{ is labeled } |0 - 4| = 4,$$
$$- e_5 = (v_0, v_5) \text{ is labeled } |0 - 5| = 5.$$

Figure 3 | Star Graph S_5 with Graceful Labeling

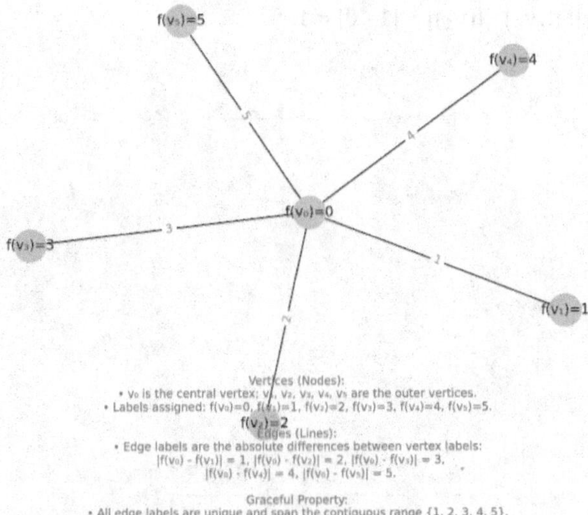

Figure 4.3 | Star Graph S5S_5S5 with Graceful Labeling

Vertices (Nodes):

- v0 is the central vertex; v1,v2,v3,v4,v5 are the outer vertices.
- Labels assigned: f(v0)=0 f(v1)=1, f(v2)=2, f(v3)=3, f(v4)=4, f(v5)=5.

Edges (Lines):

- Edge labels are the absolute differences between vertex labels:
  - |f(v0)−f(v1)|=1,
  - |f(v0)−f(v2)|=2,
  - |f(v0)−f(v3)|=3,
  - |f(v0)−f(v4)|=4,
  - |f(v0)−f(v5)|=5.

Example 2: Graceful Labeling of a Cycle Graph C_6:

A cycle graph C_6 consists of six vertices $v_1, v_2, v_3, v_4, v_5, v_6$ arranged in a closed loop with six edges connecting each consecutive pair of vertices.

Vertex Labeling: Assign vertex labels such as $f(v_1) = 0$, $f(v_2) = 5$, $f(v_3) = 2$, $f(v_4) = 3$, $f(v_5) = 4$, and $f(v_6) = 1$.

Edge Labeling: Calculate the edge labels based on the absolute differences:

- $e_1 = (v_1, v_2)$: $|f(v_1) - f(v_2)| = |0 - 5| = 5$,
- $e_2 = (v_2, v_3)$: $|f(v_2) - f(v_3)| = |5 - 2| = 3$,
- $e_3 = (v_3, v_4)$: $|f(v_3) - f(v_4)| = |2 - 3| = 1$,
- $e_4 = (v_4, v_5)$: $|f(v_4) - f(v_5)| = |3 - 4| = 1$,
- $e_5 = (v_5, v_6)$: $|f(v_5) - f(v_6)| = |4 - 1| = 3$,
- $e_6 = (v_6, v_1)$: $|f(v_6) - f(v_1)| = |1 - 0| = 1$.

Figure 4 | Cycle Graph C_6 with Graceful Labeling

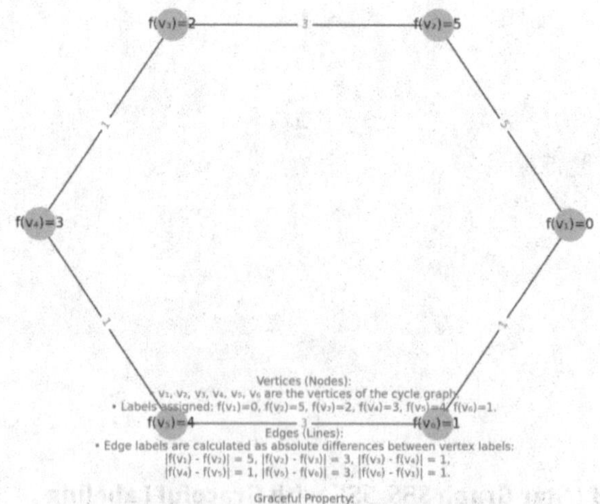

Figure 4.4 | Cycle Graph C6C_6C6 with Graceful Labeling
Vertices (Nodes):
- v1,v2,v3,v4,v5,v6 are the vertices of the cycle graph.
- Labels assigned: $f(v1)=0$, $f(v2)=5$, $f(v3)=2$, $f(v4)=3$, $f(v5)=4$, $f(v6)=1$.

Edges (Lines):
- Edge labels calculated as absolute differences between vertex labels:
  - $|f(v1)-f(v2)|=5$,
  - $|f(v2)-f(v3)|=3$,
  - $|f(v3)-f(v4)|=1$,
  - $|f(v4)-f(v5)|=1$,
  - $|f(v5)-f(v6)|=3$,
  - $|f(v6)-f(v1)|=1$.

This example demonstrates that while cycle graphs can sometimes be gracefully labeled, the requirements for uniqueness and contiguous labeling may impose constraints that prevent full graceful labeling in all cases.

4.5 Recent Advances and Computational Tools in Graceful Labeling

Graceful labeling has seen notable advancements in mutually theoretical and computational domains, which have expanded its applicability and efficiency in various fields. As research continues, new properties, tools, and techniques are developed to overcome the limitations of traditional graceful labeling. This section explores these recent advancements, computational resources that support graceful labeling, and promising future directions.

Advances in Graceful Labeling Theory

1. Newly Discovered Properties and Extensions:

Recent studies have unveiled new properties in graceful labeling that extend its theoretical foundations. These findings have deepened the understanding of how graceful labeling applied to previously unlabelable or complex graphs.

Properties of Special Graphs: For certain graph classes, researchers have identified specific properties that either support or restrict graceful labeling. For example, new criteria have been discovered for identifying which bipartite graphs can be gracefully labeled, enabling applications in broader types of bipartite networks.

Extensions to Partial and Almost Graceful Labeling: Recognizing that not all graphs can be gracefully labeled in the traditional sense, researchers have developed the concepts of partial graceful labeling and almost graceful labeling. These extensions allow for labeling subsets of vertices or edges inside a graph while approximating graceful properties. This approach has proven useful in complex graph structures, such as multipartite graphs and certain trees, where graceful labeling is challenging but partial labelability can achieve a balanced and functional labeling scheme.

2. New Labeling Techniques:

In reply for limitations of traditional graceful labeling, researchers have introduced innovative techniques that streamline the labeling process and enhance flexibility.

Modular Labeling Techniques: A recent development in modular arithmetic has enabled the use of modular sequences to achieve graceful properties in complex graphs. By applying modular arithmetic, it becomes possible to maintain unique edge labels while allowing certain degrees of flexibility in vertex labeling. This technique is especially beneficial for graphs with high degrees of connectivity.

Graph Decomposition and Subgraph Labeling: Another emerging approach involves decomposing large graphs into smaller subgraphs that can be labeled independently before combining the subgraphs back infor original structure. By labeling each subgraph gracefully, researchers can often achieve a near-graceful labeling across the entire graph, creating an effective balance in networks where strict graceful labeling is impractical.

Computational Tools for Graceful Labeling

Software Tools and Algorithms:

With the growing size and complexity of graphs in research and industry, computational tools have become indispensable for graceful labeling. These tools enable researchers to experiment with labeling strategies, verify the labelability of graphs, and manage the vast data associated with large networks.

Graph Labeling Algorithms: Specialized algorithms have been developed to assist in graceful labeling, especially for large or dense graphs where manual labeling is infeasible. Algorithms based on backtracking, branch-and-bound, and heuristic methods explore possible labeling configurations efficiently. For example, backtracking algorithms can systematically test vertex label assignments to achieve graceful properties, while branch-and-bound algorithms can reduce the total of potential configurations by eliminating unworkable options early in the process.

Software for Graph Labeling:

SageMath: SageMath is an open-source software that includes libraries for graph theory, including graph labeling. It provides functions for generating labeled graphs, performing computations, and visualizing labeling patterns. Researchers can use SageMath to test different labeling configurations and evaluate the graceful properties of their graphs.

NetworkX: NetworkX is a Python library widely expended in network analysis. Though not exclusively designed for graceful labeling, its flexibility allows users to implement graceful labeling algorithms or create custom scripts to explore labelability across various graph structures.

Graphviz: Graphviz, while primarily a visualization tool, can assist in representing gracefully labeled graphs visually. Its layout capabilities enable researchers to model complex networks and label graphs, facilitating better understanding and presentation of graceful labeling configurations.

2. Importance of Computational Tools:

Computational tools are essential for verifying graceful labelability and managing complex graph types, particularly in large-scale applications.

Verification of Graceful Labelability: For many graph classes, verifying whether a graph is gracefully labelable is a computationally intensive task. Software tools can automate this verification, saving time and improving accuracy. For example, when examining large graphs with hundreds or thousands of vertices, these tools can quickly assess feasible labelings and test for unique edge labels.

Handling Complex Graph Types: Computational tools are invaluable when trade with complex graph types, such as hypercubes or large tree structures. In these cases, automated tools simplify the labeling process by organizing data, running calculations, and producing outputs that reveal whether a given graph can meet graceful labeling conditions. This automation enables researchers to experiment with graph structures and explore potential solutions without the time-intensive burden of manual calculations.

Future Directions and Potential Applications

1. Expanding Graceful Labeling to New Graph Types: As research progresses, one major goal is to expand graceful labeling to additional graph types that are currently difficult to label gracefully. This expansion could involve developing further extensions like almost or partial graceful labeling or adapting current techniques to accommodate unique graph structures.

Focus on Complex and Non-Standard Graph Types: Future research may focus on multipartite graphs, directed acyclic graphs (DAGs), and higher-dimensional graphs. If graceful labeling can be successfully applied forse types, it would enable new applications in fields that rely on these structures, such as computational biology, logistics, and machine learning.

2.   **Emerging Applications in Key Fields:** Graceful labeling's balanced and structured properties have the potential to address challenges in arenas such as network security, quantum computing, and large-scale network simulations.

Network Security: In network security, graceful labeling could help in creating secure network topologies where balanced load distribution minimizes potential vulnerabilities. By using graceful labeling to structure communication pathways, network administrators could prevent certain links from becoming chokepoints, reducing the risk of targeted attacks.

Quantum Computing: Quantum computing is highly reliant on network structures to organize qubits and manage interactions. Graceful labeling might assist in creating configurations for quantum networks where interactions between qubits are optimized for minimal interference and balanced processing.

Large-Scale Network Simulations: In simulations for studying large networks, graceful labeling could play a role in reducing computational complexity. By ensuring an even distribution of network load, researchers can improve simulation efficiency and gain accurate insights into network behavior without the risk of overwhelming individual network nodes.

3.   **Algorithmic Development and Machine Learning Integration:** Incorporating machine learning into graceful labeling research presents another promising direction. By using machine learning models trained on various graph structures, researchers can predict graceful labelability and automate the labeling process for complex graphs.

Pattern Recognition in Labeling: Machine learning algorithms trained on labeled datasets could help identify labeling patterns and predict configurations that yield graceful properties. These algorithms could streamline the labeling process for larger graphs by recognizing commonalities across graph types.

Automated Labeling with Reinforcement Learning: Reinforcement learning techniques used to develop automated systems that learn optimal strategies for graceful labeling. By rewarding successful label assignments, reinforcement learning models can improve labeling performance over time, particularly for complex or irregular graph types.

CHAPTER 5

GRAPH COLORING

5.1 Fundamentals of Vertex and Edge Coloring

Graph coloring is a fundamental component of graph theory and has significant applications in various arenas, including computer science, scheduling, and resource allocation. The two primary types of graph coloring are vertex coloring and edge coloring, each serving a unique purpose and providing essential tools for structuring and analyzing networks.

Definition and Concepts

Vertex Coloring: Vertex coloring, process of assigning colors for vertices of a graph such that no two adjacent vertices (vertices connected by an edge) share the same color. The goal of vertex coloring is to minimalize the numeral of colors used whereas ensuring that adjacent vertices have distinct colors. This minimal number of colors needed to achieve a proper coloring is called the chromatic number for graph, often denoted by $\chi(G)$ for a graph G.

Formally, given a graph $G = (V, E)$ with vertices V and edges E, a vertex coloring is a function:

$f: V \rightarrow C$ where C is a group of colors, and for any two vertices $u, v \in V$ that are linked by an edge $(u, v) \in E$, we have $f(u) \neq f(v)$. This condition ensures that adjacent vertices do not have the same color.

Example: Consider a simple triangle graph (a cycle of three vertices). If the vertices are labeled v1, v2, v3, we could assign colors $f(v1) = 1$, $f(v2) = 2$, and $f(v3) = 1$. This coloring satisfies the conditions for vertex coloring, as each pair of adjacent vertices (connected by an edge) have different color.

Figure 5 | Vertex Coloring of a Triangle Graph

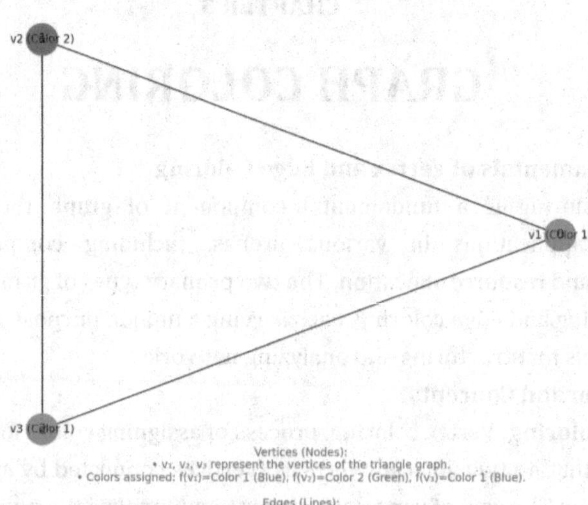

Figure 5.1 | Vertex Coloring of a Triangle Graph

Vertices (Nodes):

- v1,v2,v3v_1, v_2, v_3v1,v2,v3 represent the vertices of the triangle graph.
- Colors assigned:
  - f(v1)=Color 1 (Blue)f(v_1) = \text{Color 1 (Blue)}f(v1)=Color 1 (Blue),
  - f(v2)=Color 2 (Green)f(v_2) = \text{Color 2 (Green)}f(v2)=Color 2 (Green),
  - f(v3)=Color 1 (Blue)f(v_3) = \text{Color 1 (Blue)}f(v3)=Color 1 (Blue).

Edges (Lines):

- All edges connect vertices with different colors, satisfying the vertex coloring condition.

Edge Coloring:

Edge coloring is the process of appointing colors for edges of a graph, as no two edges occur for same vertex, and the same color. The smallest possible number of colors required for a proper edge coloring is known as a edge chromatic number, or chromatic index of the graph, often denoted by $\chi'(G)$.

Formally, for a graph G = (V, E), an edge coloring is a function:

$$g: E \rightarrow D$$

where D is the set of colors, and for any two edges e1, e2 ∈ E that share a common vertex, we have g(e1) ≠ g(e2).

Example: Consider a simple square graph (a cycle with four vertices).

Labeling the vertices v1, v2, v3, v4 and edges e1, e2, e3, e4, a valid edge coloring might assign colors g(e1) = 1, g(e2) = 2, g(e3) = 1, and g(e4) = 2. This coloring uses only two colors and ensures that no two edges event for same vertex share the same color.

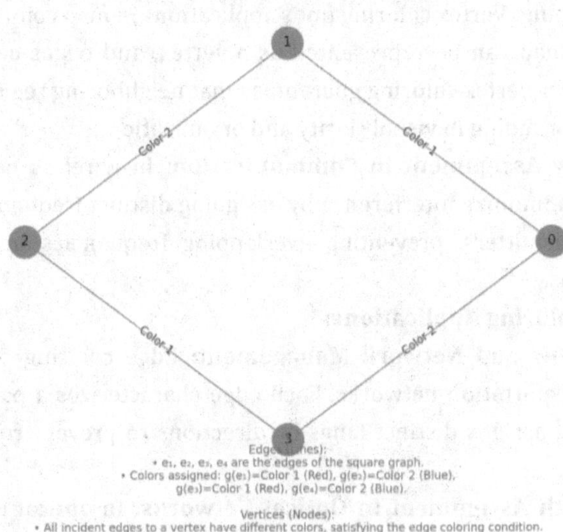

Figure 5.2 | Edge Coloring of a Square Graph

Edges (Lines):

- e1,e2,e3,e4e_1, e_2, e_3, e_4e1,e2,e3,e4 are the edges for square graph.
- Colors assigned:
- ○ g(e1)=Color 1 (Red),
- ○ g(e2)=Color 2 (Blue),
- ○ g(e3)=Color 1 (Red),
- ○ g(e4)=Color 2 (Blue).

Vertices (Nodes):

- All incident edges for vertex have different colors, satisfying the edge coloring condition.

Importance in Graph Theory

Vertex and edge coloring are foundational techniques in graph theory, as ay help solve a wide variety of practical problems by structuring complex relationships and ensuring that adjacent elements do not interfere with each other.

1. Vertex Coloring Applications:

Scheduling Problems: In scheduling, vertex coloring is used to allocate timeslots to tasks or resources without conflicts. Each task is represented as a vertex, and an edge exists between tasks that cannot occur simultaneously. Assigning colors to vertices corresponds to allocating non-overlapping timeslots.

Map Coloring: Vertex coloring finds applications in map coloring, where each region on a map can be represented as a vertex, and edges connect adjacent regions. Proper vertex coloring guarantees that neighboring regions do not have the same color, aiding in visual clarity and organization.

Frequency Assignment in Communication: In wireless networks, vertex coloring can minimize interference by assigning distinct frequencies (colors) to adjacent transmitters, preventing overlapping frequencies in geographically adjacent areas.

2. Edge Coloring Applications:

Traffic Flow and Network Management: Edge coloring can help design roads in transportation networks. Each edge characterizes a path or road, and edge coloring assigns distinct lanes or directions to prevent route conflicts at intersections.

Wavelength Assignment in Optical Networks: In optical communication, edge coloring helps assign distinct wavelengths to light signals on network paths. This process ensures that signals passing through the same junctions use different wavelengths, preventing interference and improving communication quality.

Illustrative Examples

1. Bipartite Graph:

In a bipartite graph, vertices are divided into two disjoint groups, and each edge joins a vertex in one set to a vertex in the other. Bipartite graphs are for matching problems, such as job assignments or resource allocations, where each set represents a distinct group (e.g., workers and jobs).

Vertex Coloring Example: For a bipartite graph, only two colors are needed, as are are no edges connecting vertices within the same set. For instance, in a simple bipartite graph with sets U = {u1, u2} and V = {v1, v2}, we could assign color 1 to all vertices in U and color 2 to all vertices in V.

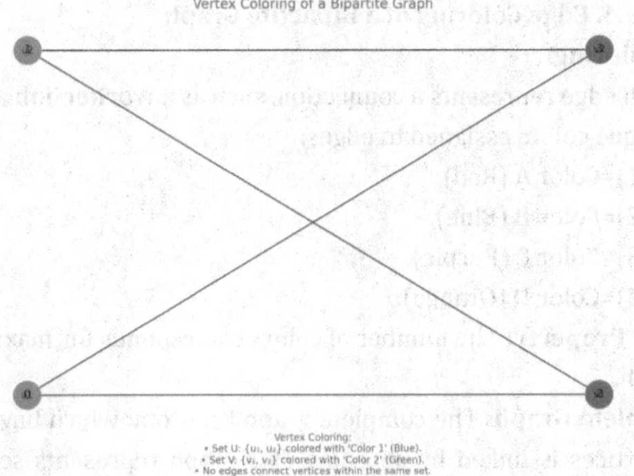

Figure 5.3. Vertex Coloring of a Bipartite Graph

Vertex Coloring:

- **Set U:** {u1,u2} colored with **Color 1 (Blue)**.
- **Set V:** {v1,v2} colored with **Color 2 (Green)**.
- **Key Property:** No edges connect vertices with same set, satisfying the bipartite graph coloring rules.

Edge Coloring Example: For a bipartite graph with high degrees, edge coloring requires at least the maximum degree of the graph. For example, in a bipartite graph representing a job assignment, each edge could be assigned a unique color representing a match between workers and jobs.

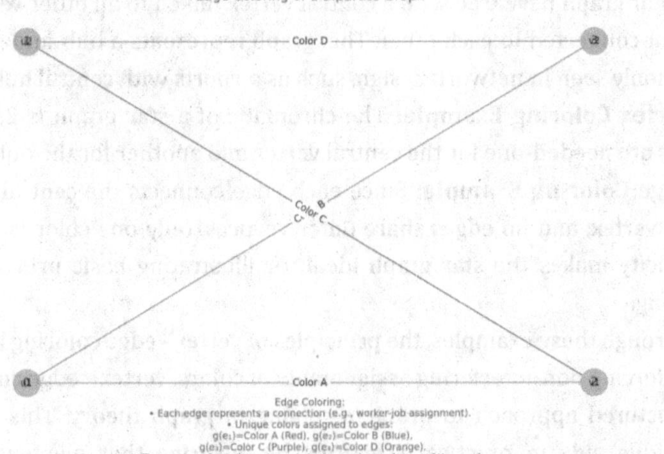

Figure 5.4. Edge Coloring of a Bipartite Graph

Edge Coloring:

- Each edge represents a connection, such as a **worker-job assignment**.
- Unique colors assigned to edges:
- g(e1)=Color A (Red),
- g(e2)=Color B (Blue),
- g(e3)=Color C (Purple),
- g(e4)=Color D (Orange).
- **Key Property:** The number of colors corresponds for maximum degree of the graph.

2. Complete Graph: The complete graph Kn is one which have every pair of distinct vertices is linked by an edge. This graph represents scenarios where every element is related to every other element, such as fully connected networks or social cliques.

Vertex Coloring Example: The chromatic of a complete graph Kn is n, as each vertex must receive a unique color to satisfy the condition of no end-to-end vertices sharing the same color. For K4 (four vertices), Four-Color s are needed, one for each vertex.

Edge Coloring Example: The chromatic index of a complete graph with an even number of vertices is n - 1, and for an odd number of vertices, it is n. For example, K4 requires three colors, as each edge connecting vertices must have a unique color that does not overlap with others at any vertex.

3. Star Graph:

A star graph have tree with a central vertex linked to all other vertices, which are not connected to each other. This graph represents a hub-and-spoke model, commonly seen in network design, such as airports with central hubs.

Vertex Coloring Example: The chromatic of a star graph is 2, as only two colors are needed-one for the central vertex and another for the outer vertices.

Edge Coloring Example: Since each edge connects the central vertex to an outer vertex, and no edges share outer vertices, only one color is needed. This simplicity makes the star graph ideal for illustrating basic principles of edge coloring.

Through these examples, the principles of vertex - edge coloring become clear. By enforcing non-interfering assignments of colors, vertex - edge coloring create a structured approach to problem-solving in graph theory. This foundational technique aids in practical scenarios by ensuring that elements like tasks, regions, and network connections are organized in a method that minimizes conflicts and optimizes efficiency.

5.2 Coloring Algorithms and Approaches

Graph coloring algorithms are essential tools in graph theory, helping solve difficulties that involve assigning colors to vertices or edges under certain constraints. This section introduces several key algorithms, discusses their complexity and efficiency, explores specialized approaches, and provides a practical example.

Overview of Common Algorithms

1. Greedy Algorithm:

The Greedy Algorithm is one that have simplest and most widely used approaches for graph coloring. This algorithm assigns colors of vertices sequentially, ensuring that each vertex receives the lowest possible color that is not already used by its adjacent vertices.

Steps:

Choose an arbitrary order of vertices.

For each vertex, assign the minimum available color that differs from the colors of adjacent vertices. Despite its simplicity, the Greedy Algorithm does not always produce the optimal (least number of colors) solution. Its performance depends heavily on the order in which vertices are chosen.

Example: Given a graph that have vertices v1, v2, v3, v4, if v1 is connected to v2 and v3, and v3 is connected to v4, the Greedy Algorithm might color the vertices as follows: v1 = 1, v2 = 2, v3 = 1, v4 = 2. This assignment uses two colors but may not be optimal.

2. Backtracking:

Backtracking is a more comprehensive approach for graph coloring, where the algorithm systematically explores all possible color assignments to find a solution that satisfies the coloring constraints.

Steps:

Assign colors for each vertex one by one, ensuring that each assignment does not violate the coloring rules.

If a conflict occurs, backtrack for previous vertex, assign a different color, and continue.

Backtracking guarantees finding an optimal solution if one exists, as it exhaustively checks all possible colorings. However, its computational complexity makes it impractical for large graphs.

3. Welsh-Powell Algorithm

The Welsh-Powell Algorithm is an enhancement over the Greedy Algorithm, primarily used to reduce the entire number of colors required in vertex coloring.

Steps:

Arrange vertices in decreasing order of their degrees (i.e., the number of edges connected to each vertex).

Apply the Greedy Algorithm, starting with the vertex with the highest degree, followed by vertices with progressively lower degrees.

By coloring high-degree vertices first, this algorithm often reduces the sum of colors needed. However, it is not always definite to find the chromatic number.

Algorithmic Complexity and Efficiency

1. Greedy Algorithm:

Complexity: O(V + E) for a graph with V vertices and E edges, as it iterates through all vertices and their neighbors.

Efficiency: The Greedy Algorithm is efficient for sparse graphs but may require significantly more colors than the optimal solution, especially when vertex ordering is suboptimal.

2. Backtracking:

Complexity: Exponential in the number for vertices, as it tries all possible color assignments. For large graphs, this is computationally expensive.

Efficiency: While computationally intensive, backtracking ensures an optimal solution, making it suitable aimed at small graphs or when an exact solution is critical.

3. Welsh-Powell Algorithm:

Complexity: O(V^2), primarily for the initial sorting step based on vertex degrees.

Efficiency: More efficient than backtracking for medium-sized graphs and often requires fewer colors than the Greedy Algorithm.

Specialized Approaches

1. Heuristic Methods:

Heuristic methods aim to find good (though not necessarily optimal) solutions quickly. These include methods such as:

Largest Degree First: Assigns colors starting from the vertex with the highest degree, similar to Welsh-Powell.

Smallest Last: Assigns colors by starting for the vertex with the fewest available colors remaining.

DSatur Algorithm: Colors the vertex with the maximum saturation (number of differently colored adjacent vertices) first.

These methods are useful in applications where approximate solutions are acceptable, and computational efficiency is a priority.

2. Probabilistic Techniques:

Probabilistic methods, like Monte Carlo simulations, use random assignments and probabilistic adjustments to find feasible colorings. They are typically applied in situations where exact solutions are not feasible or necessary, such as in large networks.

Step-by-Step Example

Consider a practical example of applying the Greedy Algorithm to color a simple graph G with vertices v1, v2, v3, v4, v5 and edges connecting adjacent vertices as follows:

- v1 is connected to v2 and v3.
- v2 is connected to v3 and v4.
- v3 is connected to v4 and v5.
- v4 is connected to v5.

We aim to color the vertices by means of the least number of colors such that no two adjacent vertices share the same color.

Steps:

1. Vertex v1: Start with the first vertex and assign it the first color, say color 1. - f(v1) = 1.

2. Vertex v2: Since v2 is adjacent to v1, it cannot have very similar color as v1. Assign it color 2. - f(v2) = 2.

3. Vertex v3: Vertex v3 is adjacent to both v1 and v2. Since these vertices already have colors 1 and 2, assign v3 color 3. - f(v3) = 3.

4. Vertex v4: Vertex v4 is adjacent to v2 and v3, so it cannot have colors 2 or 3. Assign it color 1.- f(v4) = 1.

5. Vertex v5: Vertex v5 is adjacent to v3 and v4, which have colors 3 and 1, respectively. Assign v5 color 2. - f(v5) = 2.

Final Coloring: The vertices are colored as follows:

f(v1) = 1, f(v2) = 2, f(v3) = 3, f(v4) = 1, f(v5) = 2

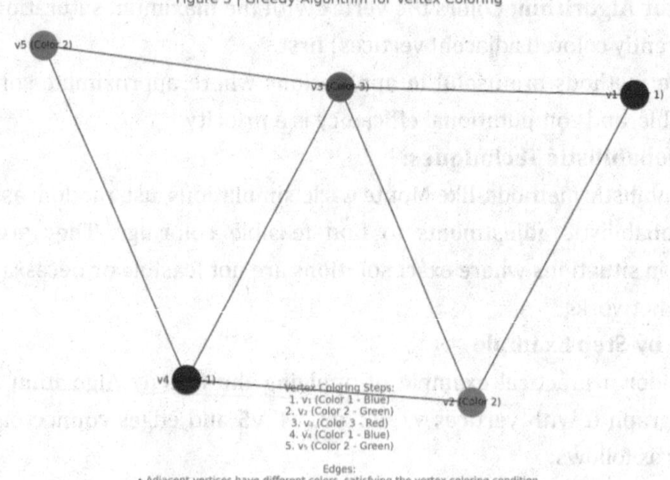

Figure 5.5 | Greedy Algorithm for Vertex Coloring

Vertex Coloring Steps:

1. v1: Assigned **Color 1 (Blue)**.
2. v2: Assigned **Color 2 (Green)**.
3. v3: Assigned **Color 3 (Red)**.
4. v4: Assigned **Color 1 (Blue)**.
5. v5: Assigned **Color 2 (Green)**.

Edges:

•   All adjacent vertices have different colors, satisfying the vertex coloring condition.

Result:

This coloring uses three colors, and no two end-to-end vertices portion the same color. While the Greedy Algorithm does not always yield the chromatic number, in this example, it provided an efficient and valid coloring for the given constraints.

Through these examples and discussions of complexity and efficiency, it is clear that choosing an appropriate graph coloring algorithm depends on the graph's structure, size, and the importance of achieving an exact solution. Greedy and heuristic algorithms provide fast, approximate solutions, while backtracking and Welsh-Powell offer pathways to more optimal colorings.

5.3 Applications in Scheduling and Allocation

Graph coloring plays a key role in scheduling and resource allocation by ensuring that conflicting tasks or resources are separated, reducing overlap, and optimizing resource use. This section discusses the presentation of graph coloring in scheduling problems, resource allocation, and network optimization, providing examples and case studies for real-world scenarios.

Scheduling Problems

Scheduling is most direct applications of graph coloring. In scheduling tasks, the goal is to allocate time slots or resources such that, the no two tasks requiring the same resource or time slot conflict. Graph coloring achieves this by ensuring that adjacent tasks (those that share a dependency or conflict) are assigned different colors (time slots or resources).

- **Timetable Creation**: Educational institutions often face the challenge of creating class schedules that avoid overlapping times for students or faculty involved in multiple courses. Each course represented as a vertex in the Graph, and an edge connects any two vertices if they share a student or teacher. Coloring the graph then corresponds to assigning time slots to each course such that no conflicting courses are scheduled for identical time.

Example: Consider four classes that scheduled in the way that avoids conflicts. Classes A and B share a student, as do B and C, and C and D. A graph with vertices representing each class and edges between classes that share students can be created. By coloring the graph, we can assign time slots to each class so no student is double-booked.

- **Project Task Scheduling**: In project management, certain tasks cannot occur simultaneously due to resource limitations or dependencies on the completion of other tasks. Representing each task as a vertex, with edges connecting tasks that cannot overlap, allows us to use graph coloring to assign time slots or resources. This approach helps optimize project timelines and ensures the availability of required resources.

Resource Allocation and Network Optimization

Graph coloring also proves useful in allocating resources within networks. In these contexts, the resources can include channels, frequencies, bandwidth, or processing units, and the factual is to minimize interference and avoid resource contention.

- **Frequency Allocation in Telecommunications**: In wireless communication networks, nearby transmitters must operate on different frequencies to avoid interference. Representing each transmitter as a vertex and drawing an edge between transmitters that are geographically close or within interference range creates a graph. Coloring this graph with a unique frequency for each transmitter ensures that adjacent transmitters do not share the same frequency, minimizing interference.

Example: Consider a network of four transmitters, each within range for the others. If transmitter A is within range of transmitters B and C, and transmitter B is within range of both transmitters A and D, a graph created where vertices represent transmitters and edges indicate potential interference. Coloring this graph allows each transmitter to operate on a frequency that minimizes interference with nearby transmitters.

- **Load Balancing in Computer Networks**: In distributed computing, graph coloring can optimize processing unit allocation across networked nodes. Tasks assigned to each processing unit need to be managed such as two tasks requiring high bandwidth or processing power overlap on the same node. By representing each task as a vertex and creating edges between tasks that cannot share the same resources, we can color the graph to allocate tasks to processing units without overloading any single unit.

Case Study: In cloud computing, multiple virtual machines (VMs) on a single server may have different levels of processing or memory requirements. Graph coloring helps allocate these VMs across servers to balance the load while avoiding conflicts between high-demand VMs.

Examples and Case Studies

1. **University Course Timetabling**:

Problem: A large university needs to create a timetable where no student is double-booked for classes, and all classes can fit within the available time slots. The university has courses that share students, meaning that overlapping these courses would cause scheduling conflicts.

Solution Using Graph Coloring: Each course is signified as a vertex, with edges between courses that share students. By coloring this graph, the university assigns time slots to each course in the way that prevents any student from being enrolled in overlapping courses.

Outcome: This approach reduces the number of time slots required and ensures an efficient, conflict-free timetable.

2. **Wireless Communication Channel Assignment**:

Problem: A telecommunications provider needs to assign frequencies to a set of cell towers positioned across a city. Towers that are geographically close to each other must operate on different frequencies to prevent signal interference.

Solution Using Graph Coloring: Each tower is signified as a vertex, and edges connect towers within interference range. Coloring the graph assigns frequencies to towers such as no two adjacent towers share the same frequency.

Outcome: The solution reduces interference across the network, allowing clear communication between towers and improving the network's overall performance and reliability.

3. **Network Task Scheduling in Data Centers**:

Problem: A data center must schedule tasks across multiple servers, ensuring that tasks with high resource demands do not run on the same server simultaneously.

Solution Using Graph Coloring: Tasks are characterized as vertices, with edges between tasks that should not share resources. By coloring the graph, tasks are assigned to servers such as high-demand tasks are distributed across multiple servers.

Outcome: This method balances the load in the data center, preventing any sole server from becoming a bottleneck and increasing the center's efficiency.

4. **Exam Scheduling in Educational Institutions**:

Problem: An institution needs to create an exam schedule where no student is required to take two exams simultaneously.

Solution Using Graph Coloring: Each exam is represented for vertex, and edges connect exams that have common students. By coloring the graph, the institution can assign different time slots to each exam in the way that avoids scheduling conflicts for students.

Outcome: The colored graph results in an efficient exam timetable that minimizes student conflicts and ensures manageable scheduling.

5. **Pipeline Management in Industrial Plants**:

Problem: An industrial plant with pipelines carrying different substances needs to avoid scheduling multiple substances through shared pipelines simultaneously, as this can lead to contamination.

Solution Using Graph Coloring: Each substance transportation process is represented as a vertex, with edges connecting processes that cannot use the pipeline simultaneously. Coloring the graph helps assign safe schedules to each process.

Outcome: The plant avoids contamination and ensures a smooth operation, optimizing pipeline use and reducing downtime.

5.4 Special Coloring Types (Chromatic and List Coloring)

In graph theory, various types of coloring help solve specific problems by addressing different constraints. Chromatic coloring and list coloring are two distinct approaches, each with unique applications, limitations, and theoretical implications.

Chromatic Coloring

Definition of Chromatic Coloring: Chromatic coloring involves assigning colors for vertices of a graph so no two adjacent vertices have the same color, similar to basic vertex coloring. However, chromatic coloring emphasizes the chromatic number for graph, denoted as $\chi(G)$, which is the least number of colors required to achieve this proper coloring for the given graph G.

The chromatic number is the intrinsic property of a graph and varies based on the structure of the graph. For example:

A bipartite graph have chromatic number for 2, as only two colors are needed to separate vertices of the two disjoint sets.

A complete graph having n vertices, denoted Kn, have chromatic number for n because every vertex must be assigned a different color.

Calculating the Chromatic Number:

Determining $\chi(G)$ can be complex, particularly for large or dense graphs. Algorithms like the Greedy Algorithm or backtracking methods used to approximate or discovery the chromatic number, although these algorithms may vary in efficiency depending on the graph's structure.

Applications and Significance of Chromatic Number:

Chromatic number has significant applications in fields requiring separation or differentiation between connected components.

Resource Allocation: In resource allocation problems, the chromatic number helps determine the minimum resources needed to avoid conflicts. For example, assigning unique channels to adjacent radio stations based on their distance or interference range.

Map Coloring: Chromatic number is used in the classic four-color theorem, which asserts that any planar map colored with no more than Four-Color s such as no two adjacent regions have the same color. This concept havepplications in geographic mapping and visual organization.

Timetabling and Scheduling: In scheduling scenarios, the chromatic number represents the minimum time slots compulsory to ensure that no overlapping tasks occur simultaneously, such as in exam scheduling where each exam shares students with others.

List Coloring

Definition of List Coloring:

List coloring is an extension of vertex coloring where each vertex for graph has its own list of permissible colors. The objective assign a color to each vertex from its specific list, ensuring that adjacent vertices do not share the same color.

Formally, in the Graph $G = (V, E)$, list coloring is achieved by defining a set of permissible colors $L(v)$ for each vertex $v \in V$. A valid list coloring is a function $f: V \to UL(v)$ such that $f(v) \in L(v)$ for all $v \in V$ and $f(u) \neq f(v)$ for all edges $(u, v) \in E$.

Applications of List Coloring:

List coloring is particularly useful when there are pre-existing constraints on the colors that assigned to vertices.

Frequency Assignment: In telecommunication networks, certain transmitters might have limited frequency options based on geographical or regulatory constraints. List coloring allows for efficient allocation while adhering forse restrictions.

Role Assignment in Team Projects: In scenarios where certain individuals are qualified only for specific tasks, list coloring ensures that each person is assigned to a role from their allowed options without conflicts with adjacent roles.

Computer Memory Allocation: In computer science, list coloring used to allocate memory segments under pre-defined constraints, ensuring efficient memory use without overlap.

Comparing Chromatic Coloring and List Coloring

While chromatic coloring and list coloring both aim to achieve a conflict-free assignment of colors, they differ in their constraints, flexibility, and feasibility in various graph structures.

Constraints:

Chromatic coloring minimizes the total number of colors used without specific color restrictions for individual vertices. In contrast, list coloring assigns colors based on predefined lists, making it more flexible but potentially more constrained if certain colors are limited.

Applications:

Chromatic coloring is widely applicable to general scheduling and allocation problems where minimizing resources is the primary objective. List coloring is used in situations where specific resources are only available to particular entities, accommodating pre-existing constraints.

Feasibility in Graph Types:

Chromatic coloring applies to all graph types but challenging to compute precisely for complex graphs. List coloring, however, may be infeasible for certain graphs if the permissible color lists are too restrictive to achieve a valid coloring.

Examples of Each Type

Example of Chromatic Coloring:

Consider a simple example with a graph G which is a triangle with three vertices v1, v2, v3, where each vertex that connected for other two. Chromatic number of this graph $\chi(G) = 3$, as each vertex needs a unique color to avoid conflicts.

Assignment: Assign colors C1, C2, and C3 to v1, v2, and v3, respectively. This ensures which pair of adjacent vertices have different color, meeting the requirements of chromatic coloring.

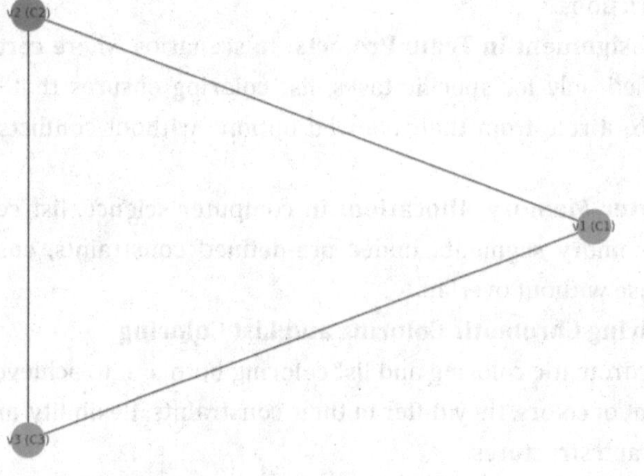

Figure 5.6. Chromatic Coloring of a Triangle Graph

Chromatic Coloring:

- v1: Assigned **C1 (Blue)**.
- v2: Assigned **C2 (Green)**.
- v3: Assigned **C3 (Red)**.

Graph Details:

- Each vertex is connected for other two, forming a triangle.
- Each vertex requires a unique color to ensure no two together vertices

share the similar color.

- **Chromatic Number χ(G)=3:** Three unique colors are necessary for proper coloring.

Example of List Coloring:

Now consider a graph H with vertices v1, v2, v3, each with its own permissible color list:

- L(v1) = {C1, C2}
- L(v2) = {C1, C3}
- L(v3) = {C2, C3}

To achieve a valid list coloring, assign colors as follows:

- Assign C1 to v1.
- Assign C3 to v2.
- Assign C2 to v3.

In this setup, each vertex is allocated a color from its permissible list, and no two adjacent vertices share the same color.

Figure 7 | Example of List Coloring

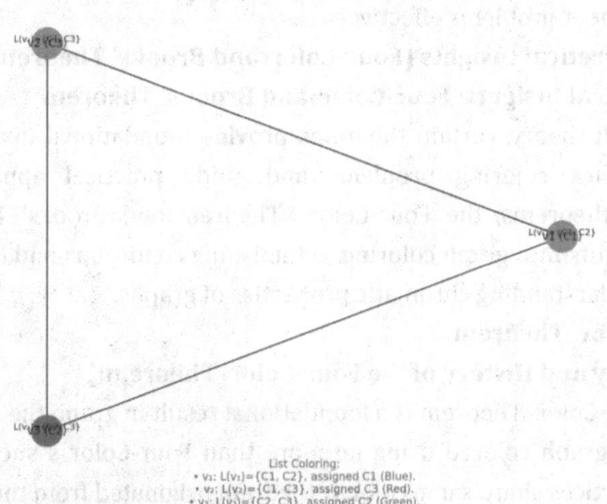

Figure 5.7 | Example of List Coloring

List Coloring:

- v1: L(v1)={C1,C2}, assigned **C1 (Blue).**
- v2: L(v2)={C1,C3}, assigned **C3 (Red).**
- v3: L(v3)={C2,C3}, assigned **C2 (Green).**

Key Properties:

- Each vertex is allocated owed a color from its permissible list.
- No two end-to-end vertices share the same color.

Case Study: Scheduling with Chromatic and List Coloring: A university needs to schedule final exams without conflict, where some exams require specific time slots based on instructor availability (list constraints), while others are flexible.

1. For exams without specific timing restrictions, chromatic coloring can control the least number of time slots required by treating each exam as a vertex and drawing edges between exams sharing students.

2. For exams with set time slots, list coloring is applied, where each exam have slant of permissible time slots. The university ensures that each student have conflict-free exam schedule while respecting instructor availability.

Through these examples and comparisons, chromatic coloring and list coloring illustrate how different coloring techniques applied to meet specific constraints and optimize resource use in graph-based scenarios. By leveraging the appropriate type of coloring, one can solve complex scheduling, allocation, and assignment problems effectively.

5.5 Theoretical Insights (Four-Color and Brooks' Theorem)

Theoretical Insights: Four-Color and Brooks' Theorem

For graph theory, certain theorems provide foundational insights that help solve complex coloring problems and guide practical applications. Two significant theorems, the Four-Color Theorem and Brooks' Theorem, offer crucial insights into graph coloring, establishing conditions and limitations that assist in understanding chromatic properties of graphs.

Four-Color Theorem

Summary and History of the Four-Color Theorem:

The Four-Color Theorem is a foundational result in graph theory that asserts any planar graph colored using no more than Four-Color s such with no two adjacent vertices share same color. It theorem originated from the map-coloring problem, which questioned the colors number required to color any political map with regions connected only at boundaries (not points) to ensure no two adjacent regions shared the same color.

The Four-Color Theorem was first speculated in 1852 by Francis Guthrie and remained unproven until 1976 when Kenneth Appel and Wolfgang Haken presented a computer-assisted proof. Their proof was first significant uses of computational methods in mathematics, as ay used a computer to verify thousands of possible configurations.

Implications for Planar Graphs: A planar graph, can be drawn on a plane without edges crossing. The Four-Color Theorem implies that every planar graph have chromatic number of at most four, meaning Four-Color s are sufficient to ensure no adjacent vertices have the same color.

The theorem has simplified many problems related to planar structures and havepplications beyond map coloring, influencing network design, scheduling, and resource allocation for any structures that represented as planar graphs.

Applications:

Geographic Mapping: In map coloring, regions represented as vertices in a planar graph, and edges represent shared boundaries. Using Four-Color s ensures that neighboring regions have distinct colors, improving map readability and aesthetics.

Network Design and Layouts: Planar networks (e.g., printed circuit boards) use four-color properties to manage connections and minimize overlaps. The theorem simplifies layouts by preventive the number of routes or colors needed.

Scheduling: Certain scheduling problems, especially those where tasks form planar dependencies, benefit from the four-color property, simplifying task arrangement without overlaps.

Brooks' Theorem

Explanation of Brooks' Theorem: Brooks' Theorem provides an upper bound on the chromatic number of a graph based on the maximum degree Δ of any vertex within the graph. The theorem states that connected graph G, the chromatic number $\chi(G)$ is at most Δ, with two exceptions:

1. If G is a complete graph Kn, then $\chi(G) = n$.
2. If G is an odd cycle, then $\chi(G) = 3$.

For most graphs, Brooks' Theorem ensures that colors number needed does not exceed the graph's maximum vertex degree. This upper bound is particularly useful when working with sparse graphs or large graphs with a predictable degree distribution.

Applications and Limitations:

Brooks' Theorem applies broadly but does not give the exact chromatic number, only an upper bound. It is often used in approximating the chromatic number for graphs not complete or odd cycles. The theorem is valuable in fields where minimizing the colors number (resources) is essential but where strict optimization is not required.

Graph Types: Brooks' Theorem is especially useful for trees, sparse graphs, and regular graphs (where all vertices with same degree). Trees, for example, have a maximum degree that provides a quick way to determine the chromatic number without detailed coloring.

Limitations: For dense graphs or graphs with high interconnectivity, Brooks' Theorem may not provide the lowest possible color count. Additionally, for complete graphs and odd cycles, Brooks' Theorem not applied as expected and requires additional coloring considerations.

Theorems in Practical Context

Both the Four-Color Theorem and Brooks' Theorem have practical applications in solving practical problems involving color assignments, resource allocation, and layout design.

1. Map Coloring and Region Distinction:

The Four-Color Theorem's principles applied to cartography and geographic mapping, ensuring that neighboring regions have distinguishable colors without redundancy. The theorem provides an efficient way to plan region colors while adhering to spatial constraints.

2. Network Design and Telecommunications:

In network design, Brooks' Theorem offers an upper bound for coloring problems where minimizing interference is crucial. By ensuring thfor chromatic number does not exceed the maximum degree, the theorem helps allocate channels or frequencies in networks with overlapping connections.

3. Scheduling in Operations Management:

For scheduling tasks with dependencies, both the Four-Color Theorem and Brooks' Theorem provide insights into reducing conflicts. The Four-Color Theorem is particularly relevant for planar scheduling problems, while Brooks' Theorem offers a guideline for tasks that form graphs with specific degrees.

Examples and Proof Outlines

Four-Color Theorem Proof Outline:

The proof of the Four-Color Theorem, developed by Appel and Haken, relied on computational verification. Here's a simplified outline of the proof concept:

1. Configuration Selection: The proof involved selecting configurations set, or graph structures, that could represent all planar graphs.

2. Reducibility: Each configuration was checked for reducibility, meaning it could be simplified without increasing the number of required colors.

3. Computer Verification: A computer program was used to exhaustively verify that every configuration could be colored by four or fewer colors.

The reliance on computational methods was groundbreaking, as it showed that some mathematical proofs could be verified only by computation, marking a shift in proof methodology.

Brooks' Theorem Proof Outline:

The proof of Brooks' Theorem is more accessible and can be outlined as follows:

Case Analysis: Brooks' Theorem is proven by analyzing various cases:

If the graph is complete graph or is odd cycle, Brooks' upper bound does not hold, as ase cases inherently require more colors.

For all other graphs, the proof demonstrates that a coloring can be achieved with Δ colors or fewer

2. Coloring Strategy: For connected graphs which not complete or odd cycles, the proof involves creating a path or ordering in which vertices are colored based on their neighbors, ensuring that no more than Δ colors are used.

Example of Four-Color Theorem: Consider a planar map divided into five regions such that each region shares boundaries with at least two other regions. By representing the map as a planar graph and using the Four-Color Theorem, we can color each region with one of Four-Color s to ensure no adjacent regions share the same color.

Example of Brooks' Theorem: Consider a tree which have maximum degree of 3. According to Brooks' Theorem, the chromatic number $\chi(G)$ of this tree is at most 3.

CHAPTER 6

HAMILTONIAN AND EULERIAN PATHS

6.1 Concepts of Hamiltonian and Eulerian Paths

In graph theory, two important types of paths are Hamiltonian and Eulerian paths. These paths are central to understanding different properties of graphs and come up with numerous applications in optimization, circuit design, and various scientific and engineering fields. This section explores the definitions, distinctions, and conditions associated with Hamiltonian and Eulerian paths and cycles.

1. Hamiltonian Path and Cycle

Definition for Hamiltonian Path:

A Hamiltonian path, graph is a path which visits each vertex of the graph for once. It may start and end at different vertices, and it does not need to cover every edge of the graph. If a Hamiltonian path exists, the graph is said to be Hamiltonian-path connected.

Definition of a Hamiltonian Cycle:

A Hamiltonian cycle (or Hamiltonian circuit) is a cycle which visits each vertex of the graph precisely once and returns for starting vertex. In this case, the path not only visits each vertex once but also completes a loop, ending for vertex where it began. A graph containing a Hamiltonian cycle is called a Hamiltonian graph.

Example of Hamiltonian Path and Cycle:

Consider a pentagon-shaped graph G with vertices V = {A, B, C, D, E} connected in a cyclic fashion with edges {(A, B), (B, C), (C, D), (D, E), (E, A)}. In this graph:

A Hamiltonian path could be A → B → C → D → E.

A Hamiltonian cycle, which returns for starting vertex, could be A → B → C → D → E → A.

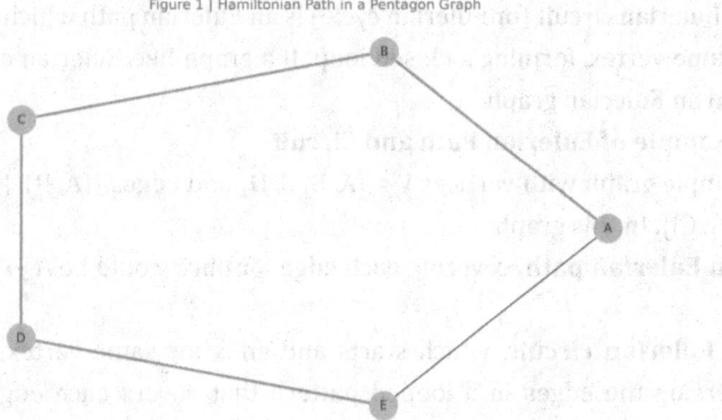
Figure 1 | Hamiltonian Path in a Pentagon Graph

Figure 6.1 | Hamiltonian Path and Cycle in a Pentagon Graph

- **Vertices:** A→B→C→D→E
- **Edges Highlighted in Red:** The path traverses separately vertex for once without forming a cycle.

Conditions for Hamiltonian Paths - Cycles:

The presence of Hamiltonian paths and cycles depends on certain structural properties of the graph. Unlike Eulerian paths - circuits, there are no simple necessary and sufficient conditions that guarantee the existence of Hamiltonian paths - cycles. However, several theorems provide sufficient conditions:

Dirac's Theorem: If graph G with n vertices (where n ≥ 3) has each vertex with degree ≥ n/2, then G contains a Hamiltonian cycle.

Ore's Theorem: If G is, graph with n ≥ 3 vertices for every pair of non-adjacent vertices u and v, the degree sum deg(u) + deg(v) ≥ n, where hen G is Hamiltonian.

These theorems provide only sufficient conditions, meaning a graph satisfying these conditions will have a Hamiltonian cycle, but a graph lacking these properties may still be Hamiltonian. For example, the Petersen graph does not meet Dirac's or Ore's criteria but contains a Hamiltonian path, illustrating thforse conditions are not necessary.

2. Eulerian Path and Circuit

Definition for Eulerian Path:

An Eulerian path in the Graph, traverses each edge for once. Unlike Hamiltonian paths, which focus on visiting vertices, Eulerian paths are concerned with covering all edges without repetition. A graph have an Eulerian path if such a path exists within it.

Definition of a Eulerian circuit:

A Eulerian circuit (or Eulerian cycle) is an Eulerian path which starts and ends for same vertex, forming a closed loop. If a graph like Eulerian circuit, then it is called an Eulerian graph.

Example of Eulerian Path and Circuit:

Simple graph with vertices V = {A, B, C, D} and edges {(A, B), (B, C), (C, D), (D, A), (A, C)}. In this graph:

An Eulerian path, covering each edge for once, could be A → B → C → A → D → C.

A Eulerian circuit, which starts and ends for same vertex, would require traversing the edges in a looped pattern that covers each edge for once and reappearances for start, such as A → B → C → A → D → C → A.

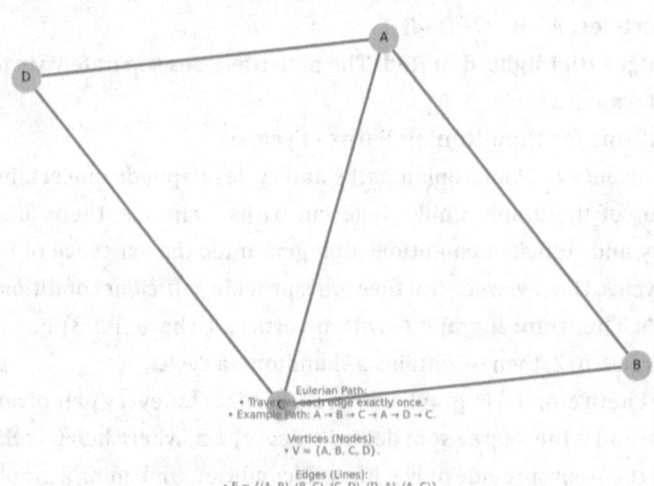

Figure 2 | Eulerian Path in a Simple Graph

Figure 6.2 | Eulerian Path for Simple Graph

Eulerian Path:

- **Vertices Traversed:** A→B→C→A→D→C
- **Edges Highlighted in Purple:** Each edge is used for once, satisfying the condition for a Eulerian path.

Graph Details:

- **Vertices (Nodes):** V={A,B,C,D}
- **Edges (Lines):** E={(A,B),(B,C),(C,D),(D,A),(A,C)}

Conditions for Eulerian Paths and Circuits:

Euler established criteria for determining whether a graph have Eulerian path or circuit:

Eulerian Circuit Condition: A connected graph have Eulerian circuit if every vertex have even degree. In other words, all vertices in the graph must have an even number of edges connected form.

Eulerian Path Condition: A connected graph have Eulerian path (but not necessarily a Eulerian circuit) if exactly two vertices have an odd degree, with the remaining vertices having an even degree. These conditions are both necessary and sufficient, meaning that any graph meeting these conditions will contain an Eulerian path or circuit. For instance, a graph which have every vertex have even degree can always form a closed loop traversing each edge once (a Eulerian circuit), while a graph with exactly two vertices of odd degree allows for an open-ended path (an Eulerian path) between those vertices.

3. Differences Between Hamiltonian and Eulerian Paths and Cycles

The primary distinctions between Hamiltonian and Eulerian paths and cycles lie in their requirements and focuses within the graph:

Vertex vs. Edge Focus:

Hamiltonian paths and cycles focus on visiting each vertex only once, regardless for number of edges.

Eulerian paths and circuits focus on traversing each edge for once, with vertex coverage being incidental to edge traversal.

Existence Conditions:

Hamiltonian paths and cycles lack simple necessary and sufficient conditions, with only sufficient conditions (such as Dirac's and Ore's theorems) providing guidance.

Eulerian paths and circuits have clear necessary and sufficient conditions based on vertex degree, which can be checked relatively easily.

Types of Problems Addressed:

Hamiltonian paths are useful in applications where visiting each location once is critical, such as a Travelling Salesman Problem.

Eulerian paths and circuits apply to problems involving the traversal of all connections or edges, such as street-sweeping and route optimization where every pathway must be covered once.

Illustrative Examples of Graphs

Example 1: Eulerian Graph Example

Consider a graph having four vertices V = {A, B, C, D} and edges {(A, B), (B, C),

(C, D), (D, A), (A, C), (B, D)}. Here, each vertex have even degree (two edges), which satisfies the conditions for a Eulerian circuit. Thus, this graph contains a Eulerian circuit, allowing traversal of each edge accurately once, starting and ending for same vertex.

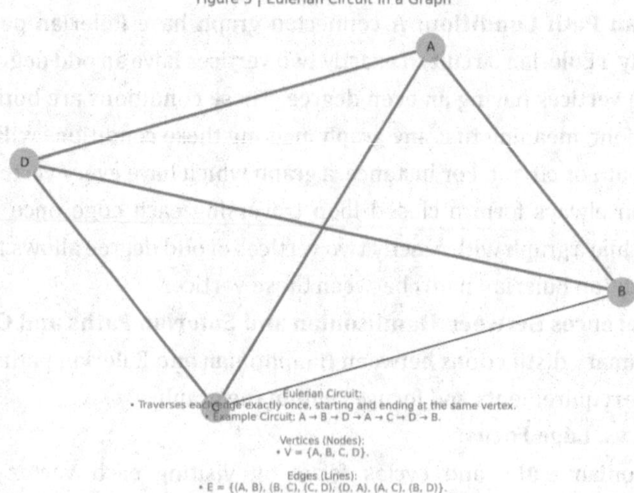

Figure 6.3 | Eulerian Circuit in the Graph

Eulerian Circuit:

- **Vertices Traversed:** A→B→D→A→C→D→B→A
- **Edges Highlighted in Green:** Each edge is traversed for once, starting and ending for same vertex, satisfying the condition for a Eulerian circuit.

Graph Details:

- **Vertices (Nodes):** V={A,B,C,D}
- **Edges (Lines):** E={(A,B),(B,C),(C,D),(D,A),(A,C),(B,D)}

Example 2: Hamiltonian Graph Example

Now consider a graph with vertices V = {1, 2, 3, 4, 5} forming a pentagon. This graph have Hamiltonian cycle that visits each vertex, before returning for start, such as 1 → 2 → 3 → 4 → 5 → 1. Since each vertex is visited only once, this path forms a Hamiltonian cycle.

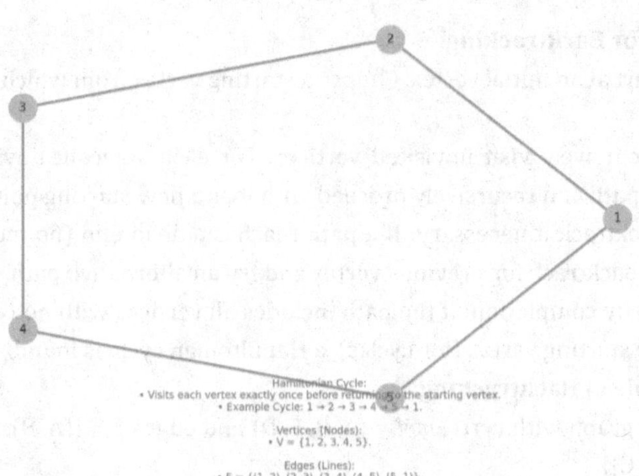

Figure 6.4 | Hamiltonian Cycle in a Pentagon Graph

Hamiltonian Cycle:

- **Vertices Traversed:** 1→2→3→4→5→1
- **Edges Highlighted in Orange:** Each vertex is visited for once before returning for starting vertex.

Graph Details:

- **Vertices (Nodes):** V={1,2,3,4,5}
- **Edges (Lines):** E={(1,2),(2,3),(3,4),(4,5),(5,1)}

6.2 Finding Hamiltonian Paths and Cycles

In graph theory, finding Hamiltonian paths and cycles involves identifying a path or cycle that visits each vertex for once in the Graph. Unlike Eulerian paths, where covering each edge for once is required, Hamiltonian paths prioritize covering all vertices without repetition. This problem have significant implications in fields like logistics and computational biology, making efficient solutions crucial. This section covers three primary methods for finding Hamiltonian paths and cycles: backtracking, dynamic programming, and heuristic approaches, along with examples to illustrate each method.

1. Backtracking Approach

Backtracking is a recursive method that travels all possible paths in the Graph, abandoning paths that fail to meet the Hamiltonian criteria. This approach is exhaustive, systematically attempting each possible route and "backtracking" when a path does not yield a Hamiltonian path or cycle. Backtracking is mainly useful for small graphs but becomes computationally prohibitive as a graph size increases.

Steps for Backtracking:

1.   Start at an initial vertex. Choose a starting vertex from which the path will begin.

2.   Recursively visit unvisited vertices: For each adjacent unvisited vertex, add it for path and recursively proceed with it as a new starting point.

3.   Backtrack if necessary: If a path reaches a dead end (no more unvisited vertices), backtrack for previous vertex and try an alternative path.

4.   Verify completion: If the path includes all vertices with no repetition and returns to starting vertex (for cycles), a Hamiltonian cycle is found.

Example of Backtracking:

Simple graph with vertices V = {A, B, C, D} and edges E = {(A, B), (B, C), (C, D), (D, A), (A, C)}.

1.   Start at vertex A.

2.   Move to B (one possible adjacent vertex).

3.   From B, move to C.

4.   From C, move to D.

5.   From D, return to A, completing a Hamiltonian cycle A → B → C → D → A.

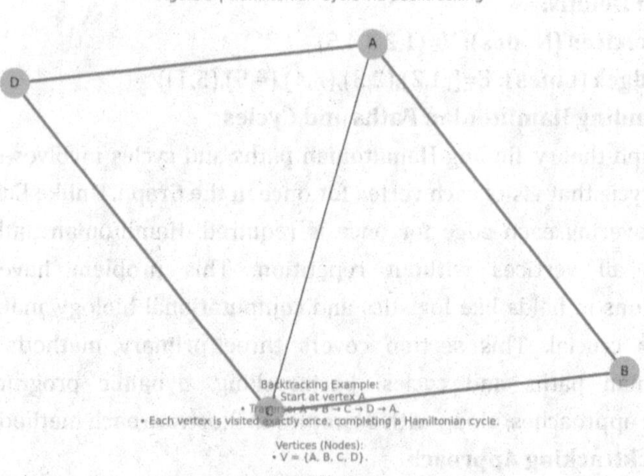

Figure 6.5 | Hamiltonian Cycle via Backtracking

Backtracking Example:

1.   **Start at Vertex:** A.

2. **Traversal Path:** A→B→C→D→A.

3. **Key Properties:** Each vertex is visited for once, completing a Hamiltonian cycle.

Graph Details:

- **Vertices (Nodes):** V={A,B,C,D}
- **Edges (Lines):** E={(A,B),(B,C),(C,D),(D,A),(A,C)}

In this approach, if any path failed to connect back to A after visiting each vertex, backtracking would occur to test alternative paths.

Advantages/Limitations of Backtracking:

Advantages: Simple and intuitive; guarantees finding a solution if one exists.

Limitations: Computationally expensive for larger graphs due to factorial complexity, making it impractical for graphs with high number of vertices.

2. Dynamic Programming Approach

Dynamic programming, optimized approach that stores results of subproblems to avoid redundant calculations. The Held-Karp algorithm, an application of dynamic programming, is commonly used for finding Hamiltonian cycles in Traveling Salesman Problem (TSP), where each city (vertex) is visited for once.

Steps in the Dynamic Programming (Held-Karp) Approach:

1. Define subproblems: Consider subsets of vertices and have track of the smallest cost of reaching each subset of vertices.

2. Store intermediate results: Use a table to record the cost (or path length) of visiting each subset of vertices.

3. Combine results: For each subset, calculate the minimum path length by adding the shortest path for current vertex and adding the previous results from smaller subsets.

4. Complete the cycle: Once all vertices are included in the subset, check paths that return for starting vertex to complete the Hamiltonian cycle.

Example of Dynamic Programming with Held-Karp:

For the graph with vertices V = {1, 2, 3, 4}, define a state for each subset and calculate the path cost incrementally.

Starting from vertex 1, find the minimum path length to visit other vertices.

Store each result in a table to ensure no path calculations are repeated.

Once all subsets are covered, add the final path back to vertex 1 for a Hamiltonian cycle.

Advantages/Limitations of Dynamic Programming:

Advantages: Reduces computational redundancy, more efficient than backtracking for certain types of graphs.

Limitations: Still has high computational complexity (typically $O(2^n * n^2)$ for the Held-Karp algorithm), making it impractical for huge graphs.

3. Heuristic Methods

Heuristic methods provide approximate solutions and are particularly useful for large graphs where exact solutions are computationally infeasible. Unlike exhaustive approaches, heuristics prioritize paths based on predefined rules, such as selecting vertices with certain properties or optimizing specific criteria (e.g., shortest paths). Common heuristic methods for Hamiltonian paths and cycles include nearest neighbor and greedy algorithms.

Nearest Neighbor Heuristic:

The next-door neighbor heuristic selects the closest (or shortest) unvisited vertex for each step, aiming to create a Hamiltonian path or cycle with minimal distance or edge count.

Steps for the Nearest Neighbor Heuristic:

1.   Start at an initial vertex.

2.   Select the neighboring unvisited vertex: Choose the nearest vertex for current vertex (according to distance or edge weight) and add it for path.

3.   Repeat until all vertices are visited: Continue selecting the adjoining unvisited vertex until the path includes all vertices.

4.   Complete the cycle (if needed): If constructing a Hamiltonian cycle, return for starting vertex.

Example of Nearest Neighbor Heuristic:

For vertices like {A, B, C, D} with distances between edges as follows:

A → B = 2

B → C = 3

C → D = 1

D → A = 4

Starting from A:

1.   Move to B (shortest path from A with a distance of 2).

2.   From B, move to C (distance of 3).

3.   From C, move to D (distance of 1).

4.   Return from D to A (completing the cycle with a distance of 4).

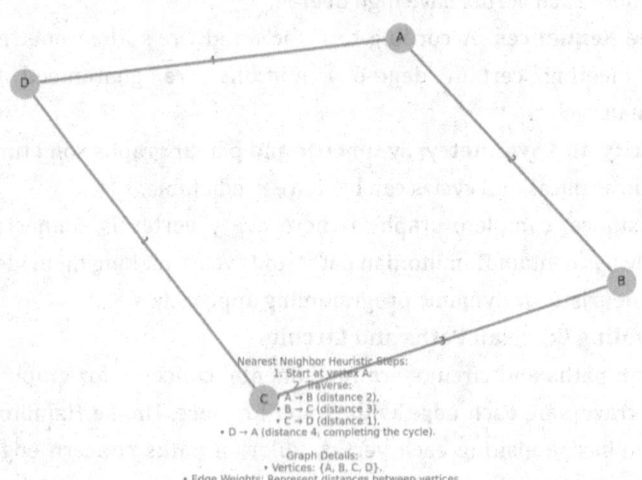

Figure 6.6 | Nearest Neighbor Heuristic Path

Nearest Neighbor Heuristic Steps:

1. **Start at Vertex:** .
2. **Traversal Path:**

o   A→B (distance 2),

o   B→C (distance 3),

o   C→D (distance 1),

o   D→A (distance 4, completing the cycle).

Graph Details:

*   **Vertices:** {A,B,C,D}
*   **Edge Weights:** Represent distances between vertices.

This heuristic might not yield the optimal Hamiltonian cycle but provides a fast, approximate solution.

Advantages/Limitations of Heuristic Methods:

Advantages: Fast and applicable to large graphs, making them suitable for practical applications where exact solutions are unnecessary.

Limitations: Does not guarantee optimality; the path may not always cover the shortest possible Hamiltonian cycle.

Special Properties Influencing Hamiltonian Paths and Cycles

Certain graph properties can simplify or complicate the search for Hamiltonian paths and cycles:

Graph Connectivity: A Hamiltonian path is more likely in highly connected graphs where each vertex have high degree.

Degree Sequences: According to Dirac's and Ore's Theorems, graphs with vertices meeting certain degree conditions are guaranteed to contain Hamiltonian cycles.

Planarity and Symmetry: Symmetric and planar graphs sometimes simplify the search, as paths and cycles can be more predictable.

For instance, complete graphs (where every vertex is connected to other vertex) always contain Hamiltonian paths and cycles, making them ideal cases for applying heuristic or dynamic programming approaches.

6.3 Finding Eulerian Paths and Circuits

Eulerian paths and circuits are fundamental concepts for graph theory that focus on traversing each edge with Graph for once. Unlike Hamiltonian paths, which emphasize visiting each vertex, Eulerian paths concern edge coverage. Euler's insights have the foundation for understanding conditions that determine the existence of such paths and circuits. This section delves into techniques and algorithms for identifying Eulerian paths and circuits, with a special on Fleury's algorithm.

1. Conditions for Eulerian Paths and Circuits

To determine if a graph contains an Eulerian path or circuit, we rely on specific conditions related to vertex degrees and graph connectivity. The degree for vertex in the Graph is like number of edges incident to it. Euler's criteria, developed in the 18th century, provide the foundational conditions for the existence of Eulerian paths and circuits.

Eulerian Circuit Condition:

A connected graph is Eulerian circuit if every vertex have even degree. This requirement ensures that each time the path enters a vertex, there is an available edge to exit without repeating any edge.

Eulerian Path Condition:

A connected graph have Eulerian path (but not necessarily a Eulerian circuit) if exactly two vertices had an odd degree. The path will start and end forse vertices.

These conditions are both necessary and sufficient, meaning that any graph meeting these conditions will contain an Eulerian path or circuit.

Example:

Consider a graph for vertices V = {A, B, C, D} and edges E = {(A, B), (B, C), (C, D), (D, A), (A, C)}.

Each vertex in this graph have even degree, satisfying the Eulerian circuit condition. Therefore, the graph contains a Eulerian circuit.

2. Fleury's Algorithm for Finding Eulerian Paths and Circuits

Fleury's algorithm is a classic approach to finding an Eulerian path or circuit in the Graph. The algorithm is straightforward and ideal for manual computation in smaller graphs. It relies on choosing edges systematically to avoid "cutting off" sections of the graph prematurely.

Steps in Fleury's Algorithm:

1. **Start at a vertex:** If there is exactly two vertices with the odd degree, start at one of these. If all vertices have an even degree, start for vertex.

2. **Select an edge to traverse:** Choose an edge from the current vertex that have not a bridge unless where is no alternative. A bridge is an edge that, if removed, would disconnect the graph.

3. **Traverse the edge and remove it:** Move for adjacent vertex along the chosen edge, then remove this edge from the graph.

4. **Repeat until no edges remain:** Continue selecting and removing edges until all edges are traversed. If the graph initially had two odd-degree vertices, this process will end for other odd-degree vertex, forming an Eulerian path. If all vertices had even degrees, the path will return for starting vertex, completing a Eulerian circuit.

Example of Fleury's Algorithm:

Consider the graph with vertices V = {A, B, C, D} and edges E = {(A, B), (B, C), (C, D), (D, A), (A, C)}, which forms a cycle with one additional edge.

1. Start at A (all vertices have an even degree, so any starting vertex is valid).

2. From A, move to B.

3. From B, move to C (avoiding A → C, which could isolate a section if removed).

4. From C, move to D.

5. Finally, from D, return to A, forming the Eulerian circuit A → B → C → D → A.

Advantages/Limitations of Fleury's Algorithm:

Advantages: Simple and easy to implement, especially for smaller graphs.

Limitations: Computationally inefficient for large graphs, as identifying bridges is time-consuming. It's more suitable for cases where manual calculations or smaller computations are involved.

3. Hierholzer's Algorithm for Finding Eulerian Circuits

Hierholzer's algorithm is specifically designed for finding Eulerian circuits in graphs where every vertex have even degree. It is more efficient than Fleury's algorithm for larger graphs, as it avoids the need to continually check for bridges. Instead, Hierholzer's algorithm constructs the circuit in stages, ensuring each edge is used for once.

Steps in Hierholzer's Algorithm:

1. Start at any vertex: Select any vertex for graph with an even degree.

2. Construct an initial cycle: Form a cycle by traversing edges and returning for starting vertex. Remove each edge as it is traversed.

3. Identify unvisited edges: If the cycle doesn't include all edges for graph, choose a vertex in the current cycle that has unvisited edges.

4. Form a subcycle and integrate it: Starting from this vertex with unvisited edges, construct another cycle that reconnects back to this vertex. Merge this subcycle with the original cycle.

5. Repeat until all edges are included: Continue creating and merging subcycles until all edges are covered. The result will be a Eulerian circuit.

Example of Hierholzer's Algorithm:

Consider a graph with vertices V = {A, B, C, D, E} and edges E = {(A, B), (B, C), (C, D), (D, E), (E, A), (A, C), (C, E)}.

1. Start at vertex A.

2. Form an initial cycle A → B → C → D → E → A.

3. Observe that vertex C has unvisited edges.

4. Starting from C, form a subcycle C → E → C and integrate this with the initial cycle.

5. The final Eulerian circuit is A → B → C → E → C → D → E → A.

Advantages/Limitations of Hierholzer's Algorithm:

Advantages: Efficient and suitable for larger graphs, as it does not require checking for bridges.

Limitations: Applicable only for finding Eulerian circuits and not paths, as it assumes all vertices has even degree.

4. Comparison of Fleury's and Hierholzer's Algorithms

While both algorithms are designed to find Eulerian paths or circuits, each is suited to different scenarios:

Fleury's Algorithm is ideal for smaller graphs or when a Eulerian path is required, as it handles cases with odd-degree vertices.

Hierholzer's Algorithm is efficient for larger graphs where a Eulerian circuit is needed, as it bypasses bridge-checking by focusing on cycles.

Applications of Hamiltonian and Eulerian Paths in Optimization and Circuit Design

Hamiltonian and Eulerian paths have profound applications for various fields, like optimization and circuit design. These paths and circuits enable efficient solutions to problems in logistics, electronics, and networking by providing structured methods for traversing graphs under specific constraints. This section explores real-world applications of these concepts, emphasizing their role in optimization and circuit design.

1. Hamiltonian Paths and Circuits in Optimization

Hamiltonian paths and cycles, which involve visiting each vertex for once, play a critical role in optimization. This is especially useful in fields where resources must be managed efficiently, such as logistics, scheduling, and routing.

Traveling Salesman Problem (TSP): One of the most famous optimization problems, TSP, requires finding the direct possible route that visits each city (vertex) for once and returns for starting point. TSP is a Hamiltonian cycle problem: it seeks a cycle in a weighted graph where the sum of the edge weights (distances) is minimized.

- **Application in Logistics and Routing:** Logistics companies use TSP-inspired algorithms to optimize delivery routes, minimizing travel time and fuel consumption. For example, a delivery company that needs to drop off packages at multiple locations within a city can use TSP algorithms to determine the optimal order of stops, thereby reducing overall distance traveled.

Example: Suppose a delivery truck has six destinations to cover in a city. By applying a Hamiltonian cycle algorithm, the company can determine an optimal path that minimizes fuel costs while ensuring each stop is visited once before returning for starting location. The impact on efficiency can be substantial, especially for large fleets covering hundreds of destinations daily.

- **Use in Manufacturing:** In manufacturing, particularly in the production line sequencing, TSP-based approaches can minimize movement between different processing stations, thus reducing time and cost. For instance, robotic arms in an assembly line might be programmed to perform tasks at multiple stations efficiently by calculating the shortest path which visits each station.

Vehicle Routing Problem (VRP): VRP is a more complex form of TSP where multiple vehicles are involved, each with capacity constraints. The problem requires optimizing routes to minimize costs while ensuring each location is visited by only one vehicle.

- **Application in Supply Chain Optimization:** In large-scale supply chain networks, companies use VRP solutions to plan routes for multiple delivery vehicles across vast geographic regions. For example, grocery delivery services use VRP-based algorithms to ensure timely deliveries by planning optimal routes that account for vehicle load, fuel costs, and delivery deadlines.

- **Case Study:** UPS developed the ORION system, an optimization tool that reduces miles driven by optimizing delivery routes. Using VRP techniques based on Hamiltonian cycles, ORION considers variables like customer locations, traffic patterns, and delivery windows, leading to fuel and cost savings across its fleet.

2. Eulerian Paths and Circuits in Optimization

Eulerian paths and circuits, involve traversing each edge for once, are particularly relevant in optimization problems where every link or connection needs coverage without redundancy. These paths are beneficial in tasks requiring thorough inspection or maintenance of connections or routes, like utility inspections and garbage collection.

Postman Problem: The Chinese Postman Problem, or Route Inspection Problem, seeks the shortest path that covers every edge of a graph for least once. This is essentially a Eulerian circuit problem.

- **Application in Urban Maintenance and Services:** Many city maintenance tasks require traversing all streets within a neighborhood without unnecessary repetition. This applies to services such as snow plowing, street sweeping, and garbage collection. By modeling city streets as edges for Graph, a service truck can use Eulerian circuits to ensure complete coverage with minimal overlap.

Example: For a city planning a street-sweeping schedule, Eulerian paths help optimize routes so that each street is swept once. If certain areas are only accessible by specific roads, an Eulerian path ensures all streets are covered while minimizing fuel and time costs.

Utility Network Inspections: Utility companies often inspect and maintain networks, such as power lines, pipelines, and telecommunications cables, which require full coverage without retracing paths unnecessarily.

- **Case Study in Power Line Inspections:** Electric companies use drones to inspect power lines, following paths that trace each line segment (edge) in the network for once. By modeling the power lines as edges and substations as vertices, companies can apply Eulerian circuit algorithms to optimize inspection routes, ensuring each line is checked without wasting drone battery life on redundant paths.

6.4 Applications in Circuit Design and Electronics

In electronics and circuit design, Hamiltonian and Eulerian paths contribute to efficient layout planning, minimizing wire lengths and improving circuit performance. Circuit design often involves creating layouts where connections must be covered efficiently, and Hamiltonian and Eulerian concepts facilitate such designs.

Hamiltonian Paths in PCB (Printed Circuit Board) Design: In PCB design, reducing the length of wiring connections between components minimizes electrical interference and power loss. Hamiltonian paths are used to optimize the layout of components on a board, where each connection needs to be visited without redundancy.

- **Application in Component Arrangement:** Hamiltonian paths can help in arranging components so that signal paths are short, minimizing latency and improving overall performance. Engineers can map circuit connections onto a Hamiltonian path to ensure each component is connected in a compact, looped layout.

Example: For a PCB that includes several chips and sensors, a Hamiltonian path can determine the optimal arrangement of components to reduce the signal travel distance. This is mainly useful in compact devices like smartphones, where space is limited, and efficiency is critical.

Eulerian Circuits in Integrated Circuit (IC) Testing: In IC testing, each node and connection on the chip must be checked for faults without covering the same path repeatedly. Eulerian circuits allow engineers to design test paths that ensure complete coverage of circuit connections, identifying potential flaws in wiring or component function.

- **Application in Scan Chain Testing:** Scan chain testing involves inspecting the connections between logic gates on a chip. By arranging the test path as a Eulerian circuit, each connection is tested once, reducing the time needed for testing.

Case Study: Intel uses Eulerian circuit principles in its scan chain testing protocols. By applying Eulerian paths, Intel ensures that each gate connection is efficiently tested for faults without rechecking the same connections, reducing the overall time for quality control and increasing the throughput of chip testing.

4. Networking Applications of Hamiltonian and Eulerian Paths

In networking, Hamiltonian and Eulerian paths help optimize the layout and function of networks, from physical layouts to data routing strategies.

Hamiltonian Paths in Data Routing and Communication Networks: In communication networks, Hamiltonian paths aid in establishing data routes that avoid redundancy. This is especially relevant in fiber-optic networks, where data needs to travel along a series of nodes (routers) efficiently.

- **Application in Peer-to-Peer (P2P) Networks:** In P2P networks, Hamiltonian paths facilitate data routing by ensuring that each node is visited in a sequence without revisiting nodes, thus minimizing latency and network congestion.

Example: BitTorrent, a P2P file-sharing protocol, applies Hamiltonian-like routing algorithms to optimize data transfer paths between peers, ensuring each peer receives data packets without bottlenecks.

Eulerian Paths in Cable and Network Infrastructure Maintenance: Telecommunication companies maintain vast networks of cables and fiber optics. An Eulerian path allows maintenance teams to cover every segment of a cable network without retracing routes, optimizing resource allocation.

- **Case Study in Fiber Optic Cable Inspections:** Companies like AT&T use Eulerian paths to plan maintenance routes for fiber optic networks. By applying Eulerian circuit principles, AT&T ensures that every segment of network is inspected while avoiding redundant coverage, which can save time and operational costs.

6.5 Key Algorithms and Problem-Solving Techniques for Hamiltonian and Eulerian Paths

In graph theory, solving for Hamiltonian and Eulerian paths requires specialized algorithms tailored to different conditions and graph structures. Key algorithms-such as Fleury's algorithm, Hierholzer's algorithm, and Hamiltonian path algorithms-each offer unique approaches for identifying these paths, each with specific strengths and limitations. This section provides a comprehensive overview of these algorithms, covering their functionality, effectiveness, complexity, and limitations.

1. Fleury's Algorithm for Eulerian Paths and Circuits

Fleury's algorithm is one of the foundational algorithms for finding Eulerian paths and circuits in the Graph. Its simplicity makes it ideal for manual computations or smaller graphs. The algorithm constructs an Eulerian path or circuit by progressively choosing edges in the way that avoids "bridges" (edges whose removal would disconnect the graph) until all edges are covered.

Steps of Fleury's Algorithm:

1. **Initial Setup:** Start at with vertex with an odd degree if finding an Eulerian path, or any the vertex if finding a Eulerian circuit.

2. **Choose an Edge:** At each step, choose an edge that is not a bridge unless there are no other available options.

3. **Traverse and Remove:** Traverse the chosen edge, moving for adjacent vertex, and remove it from the graph.

4. **Repeat Until Completion:** Continue until all edges are traversed, forming an Eulerian path or circuit.

Example of Fleury's Algorithm:

Consider a graph for vertices V = {A, B, C, D} and edges E = {(A, B), (B, C), (C, D), (D, A), (A, C)}. Starting at vertex A:

1. Choose and traverse edge A → B.

2. From B, move to C.

3. Continue from C to D, and finally return to A, completing a Eulerian circuit A → B → C → D → A.

Complexity and Limitations of Fleury's Algorithm:

Complexity: Fleury's algorithm have time complexity of $O(E^2)$ due for need to check for bridges repeatedly, which can developed computationally expensive for large graphs.

Limitations: While easy to implement, Fleury's algorithm is inefficient for large graphs which may not work well in scenarios requiring real-time calculations due to its bridge-checking requirement. It's best suited for small or manually computed graphs.

2. Hierholzer's Algorithm for Eulerian Circuits

Hierholzer's algorithm is a more efficient approach for finding Eulerian circuits, particularly in large graphs. It bypasses the need to repeatedly check for bridges by focusing on creating and merging cycles within the graph. This algorithm is applicable only for graphs where all vertex have even degree, as it specifically solves for Eulerian circuits rather than Eulerian paths.

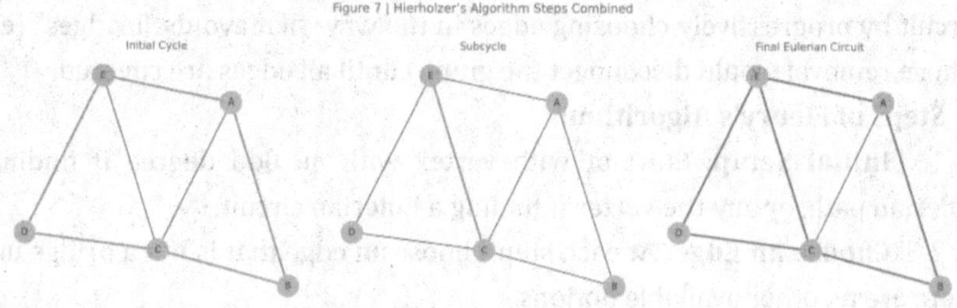

Figure 6.7 | Hierholzer's Algorithm Steps Combined

Subplots:

1. **Initial Cycle:**
○ Path: A→B→C→D→E→A (Highlighted in Blue).
2. **Subcycle:**
○ Path from vertex C: C→E→C (Highlighted in Red).
3. **Final Eulerian Circuit:**
○ Complete Circuit: A→B→C→E→C→D→E→A (Highlighted in Green).

Steps of Hierholzer's Algorithm:

1. **Start for Any Vertex:** Select any vertex for graph with an even degree.

2. **Form an Initial Cycle:** Traverse edges to form a cycle that returns for starting vertex, removing each edge as it is traversed.

3. **Identify Unvisited Edges:** If the initial cycle does not cover all edges, choose a vertex within the current cycle that still has unvisited edges.

4. **Form and Merge Subcycles:** From this vertex, form another cycle that eventually returns for starting point of the subcycle, merging it with the main cycle.

5. **Repeat Until All Edges are Covered:** Continue forming and merging subcycles until all edges in a graph are traversed, resulting in a complete Eulerian circuit.

Example of Hierholzer's Algorithm:

Consider a graph with vertices V = {A, B, C, D, E} and edges E = {(A, B), (B, C), (C, D), (D, E), (E, A), (A, C), (C, E)}.

1. Start at vertex A and form an initial cycle A → B → C → D → E → A.

2. Notice that vertex C has unvisited edges.

3. Form a subcycle C → E → C and integrate it with the initial cycle.

4. The final Eulerian circuit is A → B → C → E → C → D → E → A.

Complexity and Limitations of Hierholzer's Algorithm:

Complexity: Hierholzer's algorithm have time complexity of O(E), as each edge is traversed for once, making it highly efficient for large graphs.

Limitations: Hierholzer's algorithm is limited to Eulerian circuits and requires that all vertices has even degree, so it cannot find Eulerian paths in graphs with odd-degree vertices.

3. Hamiltonian Path Algorithms

Hamiltonian path algorithms differ from Eulerian algorithms in thfory focus on visiting each vertex for once rather than covering each edge. There is no single efficient solution for finding Hamiltonian paths or cycles, as problems are NP-complete. However, various approaches can be applied depending on exact requirements and graph size.

Backtracking Approach:

Backtracking is a brute-force approach for finding Hamiltonian paths and cycles by exhaustively exploring possible vertex sequences. This method is straightforward but computationally expensive, as it evaluates all permutations of vertices.

Steps of the Backtracking Approach:

1. Start from an Initial Vertex: Select a starting vertex and initiate a recursive search.

2. Visit Adjacent Unvisited Vertices: For each adjacent unvisited vertex, add it for current path.

3. Check for Completion: If all vertices visited, check if there will be an edge back for starting vertex (for Hamiltonian cycles).

5. Backtrack if Necessary: If the current path fails to cover all vertices or does not form a cycle, backtrack to explore alternative paths.

Example of Backtracking for a Hamiltonian Path:

For a graph which have vertices V = {1, 2, 3, 4} and edges that connect each vertex:

1. Start at vertex 1.

2. Move to vertex 2, then to vertex 3, and finally to vertex 4.

3. If all vertices are covered and there is a return path for starting vertex, a Hamiltonian cycle is formed.

Complexity and Limitations for Backtracking Approach:

Complexity: This approach have factorial time complexity of O(n!), making it unreasonable for large graphs.

Limitations: Backtracking is computationally infeasible for large graphs but can be useful for small graphs or in educational contexts.

Dynamic Programming Approach (Held-Karp Algorithm)

The Held-Karp algorithm is which have dynamic programming solution for finding Hamiltonian cycles, commonly applied in Traveling Salesman Problem (TSP). By storing results of subproblems, this approach reduces redundant calculations, although it remains computationally intensive.

Steps of the Held-Karp Algorithm:

1.   **Define Subproblems:** For subsets of vertices, calculate the minimum cost to reach each subset.

2.   **Store Intermediate Results:** Use a table to record path costs for each subset.

3.   **Combine Results:** For each subset, calculate the minimum path cost by combining previously calculated values.

4.   **Complete the Cycle:** Once all vertices are included, add the path back for starting vertex for a Hamiltonian cycle.

Example Using Held-Karp:

For a weighted graph vertices V = {1, 2, 3, 4}:

1.   Define subsets such as {1, 2}, {1, 3}, and calculate minimum path costs incrementally.

2.   Store each result in a table to avoid recalculating paths.

3.   Once all vertices are covered, calculate the minimal Hamiltonian cycle.

Complexity and Limitations of Held-Karp Algorithm:

Complexity: The Held-Karp algorithm has complexity of $O(2^n * n^2)$, which is more efficient than backtracking but still challenging for very large graphs.

Limitations: While it improves on brute-force approaches, the Held-Karp algorithm remains impractical for graphs with the high number of vertices due to its exponential complexity.

Summary of Algorithm Effectiveness

Fleury's Algorithm: Ideal for finding Eulerian paths in smaller graphs; limited by the need for bridge-checking.

Hierholzer's Algorithm: Efficient for Eulerian circuits in large graphs, bypassing bridge-checking requirements.

Backtracking (Hamiltonian Paths): Useful for small graphs; computationally intensive for larger ones.

Held-Karp (Dynamic Programming): Suitable for Hamiltonian cycles in medium-sized graphs, balancing accuracy with computational limits.

CHAPTER 7

GRAPH ISOMORPHISM

7.1 Introduction to Graph Isomorphism

In graph theory, graph isomorphism that is fundamental concept that examines whether two graphs have structurally identical, even if their representations may differ. Isomorphism allows us to understand when two graphs are "the same" in a structural sense, providing essential insights in fields such as chemistry, computer science, and network analysis. This section provides a detailed explanation of graph isomorphism, the criteria for identifying isomorphic graphs, and practical applications.

1. Definition of Graph Isomorphism

Two graphs $G_1 = (V_1, E_1)$ and $G_2 = (V_2, E_2)$ are considered isomorphic if there exists a one-to-one correspondence between their vertices and edges that preserves the adjacency relationships. In other words, two graphs are isomorphic if they have the same structure, but the labels or names of their vertices and edges may differ.

Formal Definition:

Graphs $G_1 = (V_1, E_1)$ and $G_2 = (V_2, E_2)$ are isomorphic if there exists a bijective function f: $V_1 \rightarrow V_2$ such that:

For every edge $(u, v) \in E_1$, the corresponding pair $(f(u), f(v)) \in E_2$.

This bijection f is called an isomorphism between G_1 and G_2. The existence of an isomorphism implies thfor two graphs are identical in terms of their connectivity patterns and overall structure.

Example: Consider the following two graphs:

Graph G_1: Vertices $V_1 = \{A, B, C\}$ with edges $E_1 = \{(A, B), (B, C), (C, A)\}$.

Graph G_2: Vertices $V_2 = \{X, Y, Z\}$ with edges $E_2 = \{(X, Y), (Y, Z), (Z, X)\}$.

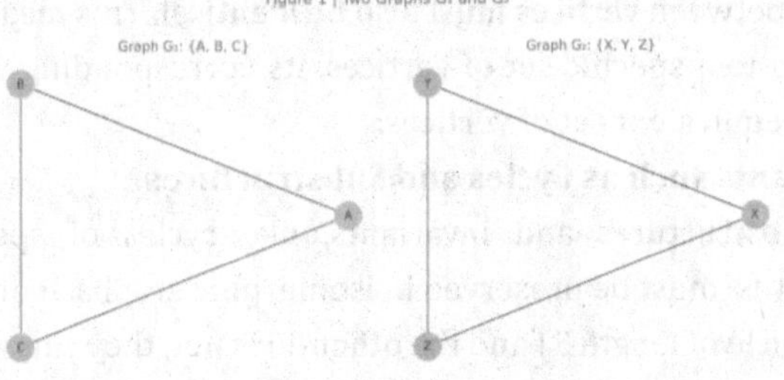

Figure 1 | Two Graphs G₁ and G₂

Figure 7.1 | Two Graphs G_1 and G_2

Graph G_1:

- **Vertices:** V1={A,B,C}
- **Edges:** E1={(A,B),(B,C),(C,A)}

Graph G_2:

- **Vertices:** V2={X,Y,Z}
- **Edges:** E2={(X,Y),(Y,Z),(Z,X)}

These graphs are isomorphic because there exists a mapping f where f(A) = X, f(B) = Y, and f(C) = Z, preserving adjacency. Each edge in G_1 have corresponding edge in G_2, meaning the two graphs share the same structural properties despite differences in vertex labels.

2. Conditions for Graph Isomorphism

To determine whether two graphs are isomorphic, certain criteria must be met. These conditions are necessary but not always sufficient on their own to confirm isomorphism; however, they can help identify non-isomorphic pairs of graphs efficiently.

1. **Equal Number of Vertices and Edges:**

For two graphs to be isomorphic, they must have the same number of vertices and edges. If G_1 and G_2 differ in the number of vertices or edges, they cannot be isomorphic.

2. **Degree Sequence Matching:**

The degree sequence of a graph is the list of vertex degrees arranged in non-increasing order. For two graphs to be isomorphic, their degree sequences must match. The degree sequence provides a necessary condition, as vertices with different degrees cannot correspond to each other in an isomorphism.

Example: If one graph have vertex of degree 3 and the other graph has no vertex of degree 3, they cannot be isomorphic.

3. **Preserved Connectivity Patterns:**

Beyond matching degree sequences, the connectivity or adjacency relationships between vertices must also be identical. This means that if a vertex in G_1 connects to a specific set of vertices, its corresponding vertex in G_2 must connect to an equivalent set of vertices.

4. **Invariants such as Cycles and Substructures:**

Certain substructures and invariants, like cycles of specific lengths or particular motifs, must be preserved in isomorphic graphs. If one graph contains a triangle (a cycle of length 3) and the other does not, they cannot be isomorphic.

Similarly, if one graph contains specific cliques or cut vertices, the other must contain equivalent structures.

These conditions help identify non-isomorphic graphs quickly, though proving isomorphism still requires a bijective mapping that confirms all adjacency relationships are preserved.

3. Examples of Isomorphic and Non-Isomorphic Graphs

Example 1: Isomorphic Graphs

Consider two graphs, G_3 and G_4:

Graph G_3: Vertices $V_3 = \{1, 2, 3, 4\}$ with edges $E_3 = \{(1, 2), (2, 3), (3, 4), (4, 1)\}$.

Graph G_4: Vertices $V_4 = \{A, B, C, D\}$ with edges $E_4 = \{(A, B), (B, C), (C, D), (D, A)\}$.

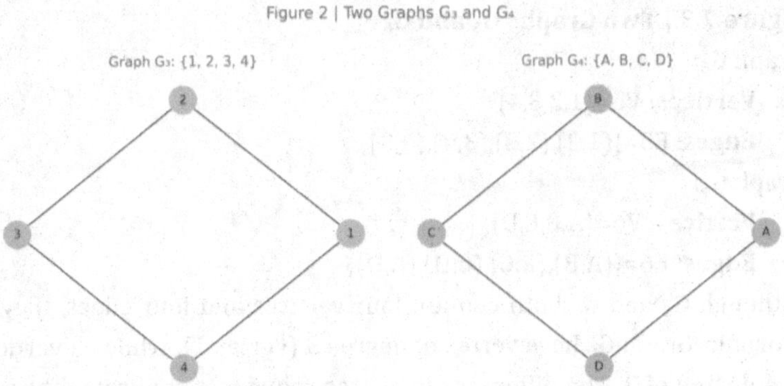

Figure 2 | Two Graphs G_3 and G_4

Graph G_3: {1, 2, 3, 4}   Graph G_4: {A, B, C, D}

Figure 7.2 | Two Graphs G_3 and G_4

Graph G_3:

- **Vertices:** V3={1,2,3,4}
- **Edges:** E3={(1,2),(2,3),(3,4),(4,1)}

Graph G4:

- **Vertices:** V4={A,B,C,D}
- **Edges:** E4={(A,B),(B,C),(C,D),(D,A)}

These graphs are isomorphic because they each represent a four-vertex cycle. A mapping such as f(1) = A, f(2) = B, f(3) = C, f(4) = D preserves all adjacency relationships, making them structurally identical.

Example 2: Non-Isomorphic Graphs

Consider two graphs G_5 and G_6:

Graph G_5: Vertices $V_5 = \{1, 2, 3, 4\}$ with edges $E_5 = \{(1, 2), (2, 3), (3, 4), (1, 3)\}$.

Graph G_6: Vertices $V_6 = \{A, B, C, D\}$ with edges $E_6 = \{(A, B), (B, C), (C, D), (A,$

D)}}.

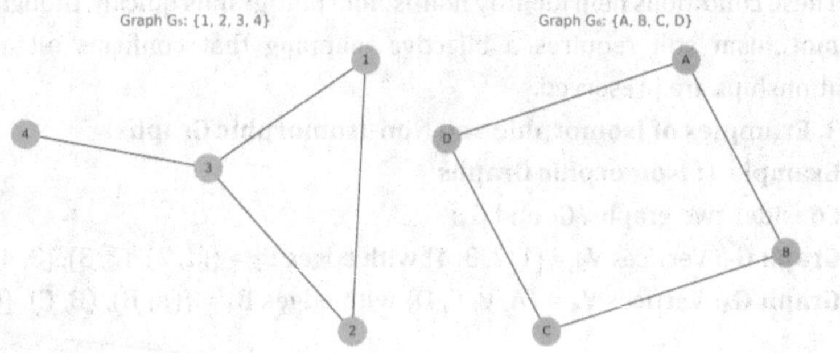

Figure 7.3 | Two Graphs G_5 and G_6

Graph G_5:

- **Vertices:** V5={1,2,3,4}
- **Edges:** E5={(1,2),(2,3),(3,4),(1,3)}

Graph G_6:

- **Vertices:** V6={A,B,C,D}
- **Edges:** E6={(A,B),(B,C),(C,D),(A,D)}

Although G_5 and G_6 both contain four vertices and four edges, they are not isomorphic. Graph G_5 have vertex of degree 3 (vertex 1), while all vertices in G_6 have a degree of 2. The difference in degree sequences confirms thforse graphs are non-isomorphic.

4. Practical Applications of Graph Isomorphism

Graph isomorphism is crucial in many practical applications where structural similarity or equivalence is essential. Below are some notable fields where graph isomorphism plays a vital role:

1. Chemistry and Molecular Structure Identification:

In chemistry, molecules can be represented as graphs where atoms are vertices and bonds are edges. Graph isomorphism helps chemists identify molecules that share the same structure but may be represented differently. By identifying isomorphic graphs, chemists can recognize equivalent molecular structures or detect cases of stereoisomerism.

Example: Benzene (C_6H_6) can be represented as a cyclic graph with six vertices and six edges. Different visual representations of benzene are isomorphic, as ay reflect the same chemical structure.

2. Pattern Recognition and Image Processing:

Graph isomorphism is used in pattern recognition to match substructures within complex images. For instance, in facial recognition, the structure of key facial features (eyes, nose, mouth) can be represented as a graph. Matching an input graph (face) with stored templates using isomorphism allows for recognition despite variations in angle, scale, or lighting.

Example: In fingerprint analysis, minutiae points (characteristic features) can be represented as vertices, with connections indicating relative positions. Graph isomorphism helps match these structures between different fingerprint samples.

3. Social Network Analysis:

In social network analysis, users are vertices, and connections (friendships, interactions) are edges. Graph isomorphism is used to detect similar sub-networks or patterns across different networks, helping to identify communities, influential groups, or recurring interaction patterns.

Example: Detecting friend circles in a social network, represented as cliques, involves finding isomorphic subgraphs where each vertex is interconnected, revealing tightly-knit groups of individuals.

4. Database and Information Retrieval:

In databases, especially those managing complex relationships like biological pathways or protein-protein interactions, graph isomorphism helps in retrieving structurally similar entries. Identifying isomorphic structures within a database can simplify search and retrieval processes by grouping equivalent entries.

Example: In a biological database, protein interaction networks can be represented as graphs. Identifying isomorphic graphs can help find proteins with similar interaction patterns, aiding in function prediction.

7.2 Algorithms for Testing Graph Isomorphism

Testing whether two graphs are isomorphic involves finding a bijective mapping between the vertices of the graphs that preserves adjacency relationships. While graph isomorphism is straightforward to understand conceptually, testing it computationally is challenging, particularly for large and complex graphs. Several algorithms have been developed for this purpose, each with unique approaches, strengths, and limitations. This section discusses three prominent algorithms for graph isomorphism testing: the Weisfeiler-Lehman algorithm, Ullmann's algorithm, and the VF2 algorithm, along with examples to illustrate each approach.

1. Weisfeiler-Lehman Algorithm

The Weisfeiler-Lehman (WL) algorithm, often called the WL test, is a powerful heuristic algorithm for distinguishing non-isomorphic graphs. The WL algorithm, especially its k-dimensional version, has been widely applied in machine learning for graph data due to its efficiency and effectiveness. This algorithm operates by iteratively refining vertex labels based on neighboring vertices, creating unique vertex "signatures" that help differentiate graphs.

Steps of the Weisfeiler-Lehman Algorithm:

1. Initial Labeling: Assign each vertex a label. Initially, labels may be based on vertex degree or set to a default value.

2. Neighbor Aggregation: For each vertex, aggregate labels from neighboring vertices to create a new signature for that vertex. This step captures the local structure around each vertex.

3. Label Compression: Assign a unique label to each unique signature generated. This step refines the graph, distinguishing vertices based on neighborhood structures.

4. Iteration: Repefor aggregation and compression steps until labels no longer change. This process generates a stable label for each vertex, known as a WL label.

5. Comparison: Compare the WL labels of the two graphs. If the WL label sequences match, the graphs may be isomorphic (further verification may be needed), while mismatched labels guarantee non-isomorphism.

Example of the Weisfeiler-Lehman Algorithm:

Consider two graphs G_1 and G_2, each with vertices labeled according forir degrees. In each iteration, the WL algorithm refines the labels based on neighboring vertices. For instance:

After initial labeling, suppose vertex labels in G_1 and G_2 are identical.

During the next iteration, if neighborhood labels differ, the WL algorithm will produce different labels for vertices in G_1 and G_2, confirming non-isomorphism.

Strengths and Applications:

The WL algorithm is efficient and can distinguish a large variety of non-isomorphic graphs.

It is especially effective for graphs with unique local structures, making it useful in machine learning, where graph data must be processed quickly.

Limitations:

The WL algorithm is not a complete test; it may fail for certain graph pairs, particularly strongly regular graphs, which require additional testing to confirm isomorphism.

2. Ullmann's Algorithm

Ullmann's algorithm is a backtracking-based approach for testing graph isomorphism. It constructs a search tree, attempting to map each vertex in one graph to a vertex in the other while preserving adjacency. Ullmann's algorithm uses matrix operations and a pruning technique that significantly reduces the search space, making it suitable for relatively small graphs or graphs with distinct structure.

Steps of Ullmann's Algorithm:

1.   Adjacency Matrix Representation: Represent both graphs as adjacency matrices A_1 and A_2.

2.   Initial Mapping Matrix: Create a mapping matrix M with entries set to 1 where potential vertex mappings could exist and 0 otherwise.

3.   Recursive Backtracking: For each vertex in G_1, map it to a vertex in G_2 and check if this mapping preserves adjacency using the adjacency matrices. If the mapping violates adjacency, backtrack for previous step.

4.   Pruning the Search Space: Use the mapping matrix to exclude invalid mappings as a algorithm progresses. This pruning helps limit the number of possible mappings.

5.   Isomorphism Verification: If a complete mapping that preserves adjacency is found, the graphs are isomorphic. If no such mapping exists after exhausting the search tree, the graphs are not isomorphic.

Example of Ullmann's Algorithm:

Consider two graphs G_3 and G_4 with four vertices each. Represent G_3 and G_4 as adjacency matrices:

Matrix A_3 (for G_3): Shows connections between vertices in G_3.

Matrix A_4 (for G_4): Shows connections in G_4.

Using Ullmann's algorithm, we attempt to map each vertex in G_3 to a vertex in G_4 while ensuring that adjacency relationships in A_3 correspond to those in A_4. By pruning and backtracking, we reduce the possible mappings until a complete, valid mapping is found.

Strengths and Applications:

Ullmann's algorithm is straightforward and systematically checks all possible mappings, ensuring accuracy.

It is effective for smaller graphs or when structural distinctions simplify the search.

Limitations:

Ullmann's algorithm becomes computationally expensive as a size of the graph increases due to its factorial time complexity.

The backtracking process limits its practicality for large or dense graphs, as it requires significant computational resources.

3. VF2 Algorithm

The VF2 algorithm is a more advanced, widely used algorithm for both graph isomorphism and subgraph isomorphism testing. VF2 builds on the principles of Ullmann's algorithm but includes sophisticated data structures and an efficient matching strategy, making it effective for larger and more complex graphs. VF2 is known for its state-space representation and ability to handle partial mappings dynamically.

Steps of the VF2 Algorithm:

1. **Initial State Setup:** Begin with an empty mapping between the two graphs.

2. **Extend Mapping:** Incrementally build a mapping by selecting vertices from each graph. For each selected pair of vertices, check compatibility based on adjacency.

3. **Feasibility Check:** At each step, use feasibility rules to ensure that mapped vertices preserve graph structure, including adjacency and neighborhood consistency.

4. **Backtracking with Pruning:** If an inconsistency is found, backtrack and prune invalid mappings from the search space.

5. **Completion:** If a complete mapping is achieved, the graphs are isomorphic. If the algorithm exhausts all possible mappings without finding a complete one, the graphs are non-isomorphic.

Example of the VF2 Algorithm:

Let G_5 and G_6 be two graphs with five vertices each. The VF2 algorithm:

Begins with an empty mapping and iteratively selects compatible vertex pairs from G_5 and G_6.

As it maps each pair, it verifies that mapped vertices retain neighborhood consistency and adjacency.

If at any step a mapping fails the feasibility check, the algorithm backtracks and prunes the search space, leading to an efficient traversal of possible mappings.

Strengths and Applications:

VF2 is highly efficient and capable of handling larger graphs than Ullmann's algorithm, owing to its sophisticated pruning and feasibility checks.

It is widely used in applications requiring graph matching, such as bioinformatics, computer vision, and database management.

Limitations:

Although efficient, VF2 may still encounter challenges with highly complex or dense graphs, as it relies on recursive backtracking.

The algorithm's performance can be impacted by the graph's structure, with denser graphs generally requiring more computational effort.

7.3 Complexity and Special Cases of Graph Isomorphism

The graph isomorphism problem presents a unique challenge in computational complexity theory, primarily due to its ambiguous placement in complexity classes. Unlike many well-defined problems in NP, graph isomorphism does not clearly fall into either the P or NP-complete categories, positioning it as an "NP-intermediate" problem. This section explores the complexity of graph isomorphism, examining why it remains a distinct issue in complexity theory. Additionally, it discusses special cases where the problem becomes computationally simpler, including examples and known algorithms that provide efficient solutions in these cases.

1. Complexity of the Graph Isomorphism Problem

The Graph Isomorphism Problem: Given two graphs $G_1 = (V_1, E_1)$ and $G_2 = (V_2, E_2)$, the graph isomorphism problem asks whether there exists a bijective mapping between V_1 and V_2 that preserves adjacency relationships. Solving this problem requires identifying an isomorphism function that maintains the structure of the graphs, which can be computationally challenging as graph size and complexity increase.

Complexity Class and NP-Intermediate Status:

Graph isomorphism is in NP because any solution (an isomorphism mapping) can be verified in polynomial time. However, it is not classified as NP-complete, as no polynomial-time reduction from any NP-complete problem to graph isomorphism has been discovered. Nor is it confirmed to be in P, as no polynomial-time algorithm that universally solves graph isomorphism has been found. This has placed graph isomorphism in a special class known as NP-intermediate, a category hypothesized by Richard Ladner for problems neither in P nor NP-complete.

Recent Developments:

In 2015, Laszlo Babai introduced an algorithm with quasi-polynomial time complexity for general graph isomorphism, making significant progress toward a potential polynomial-time solution. Babai's algorithm runs in $O(2^{(\log n)^c})$ time, where n is the number of vertices, and c is a constant. This approach has substantially narrowed the gap between quasi-polynomial and polynomial complexity for graph isomorphism, though it has yet to confirm a conclusive classification for the problem.

2. Special Cases of Graph Isomorphism

While the general graph isomorphism problem remains computationally challenging, several classes of graphs allow for efficient isomorphism testing. These include trees, planar graphs, and bounded-degree graphs, each of which has unique structural properties that simplify isomorphism testing.

A. Trees

Trees are a special type of graph characterized by their acyclic nature and hierarchical structure. A tree with n vertices has exactly n - 1 edges and no cycles, properties that reduce the complexity of isomorphism testing significantly.

Tree Isomorphism Testing:

Tree isomorphism can be solved in linear time using a canonical form approach. This method involves assigning unique labels to vertices by examining the tree's structure from the leaves inward. By recursively labeling and comparing the structure, we can determine if two trees are isomorphic.

Example of Tree Isomorphism:

Consider two trees T_1 and T_2, each with five vertices connected in a simple chain. By labeling each node starting from the leaves and moving toward the root, we create a unique string for each tree based on its hierarchical structure. If these strings match, the trees are isomorphic; otherwise, they are not.

Known Algorithm:

A widely used algorithm for tree isomorphism is based on Prüfer sequences or rooted tree codes, both of which represent trees as unique identifiers. A linear-time algorithm, such as that proposed by Aho, Hopcroft, and Ullman, effectively solves tree isomorphism by labeling the tree nodes and constructing a canonical form.

B. Planar Graphs

Planar graphs can be drawn on a plane without edges crossing, a property that facilitates isomorphism testing. The structure of planar graphs provides topological constraints that simplify isomorphism testing relative to general graphs.

Planar Graph Isomorphism Testing:

Planar graph isomorphism is solvable in polynomial time using algorithms that exploit the graph's planar embedding. For instance, planar graphs can be tested for isomorphism by encoding the embedding information into a canonical form, ensuring any two isomorphic planar graphs yield identical embeddings.

Example of Planar Graph Isomorphism:

Suppose we have two graphs representing road networks in separate cities, each designed to be planar (without intersections). By analyzing their planar embeddings and boundary cycles, we can encode each network into a unique form. If these forms match, the graphs are isomorphic; if not, they are non-isomorphic.

Known Algorithm:

The Hopcroft and Tarjan algorithm, developed in the 1970s, can test planar graph isomorphism efficiently by leveraging the constraints imposed by planar embeddings. Their algorithm is polynomial-time, and further optimizations have reduced its complexity in specific planar cases.

C. Bounded-Degree Graphs

Bounded-degree graphs are graphs where the degree of each vertex is limited to a fixed constant. For example, in a degree-3 bounded graph, no vertex connects to more than three others. This restriction significantly reduces the number of potential mappings, making isomorphism testing simpler.

Bounded-Degree Graph Isomorphism Testing:

For graphs with a bounded maximum degree, the isomorphism problem is solvable in polynomial time due for constrained nature of vertex mappings. The reduced complexity comes from the limited possibilities for each vertex's neighbors, allowing efficient backtracking-based algorithms.

Example of Bounded-Degree Graph Isomorphism:

Consider two chemical compounds, each represented by a graph where each atom (vertex) forms a maximum of three bonds (edges). With the degree bounded by three, the number of possible isomorphisms is manageable, allowing rapid testing.

Known Algorithm:

A specialized approach that applies branching and backtracking works efficiently for bounded-degree graphs. Babai and Luks developed an algorithm that leverages this structure, reducing the complexity to polynomial time for graphs with bounded degrees.

3. Additional Special Cases and Efficient Algorithms

Beyond trees, planar graphs, and bounded-degree graphs, several other graph classes permit efficient isomorphism testing. These cases involve specific structural constraints that streamline the mapping process.

Strongly Regular Graphs

A strongly regular graph is one where each vertex has a same number of neighbors, and any two adjacent vertices share a fixed number of common neighbors. This uniformity can simplify isomorphism testing by allowing for symmetry-based pruning in search algorithms.

Known Algorithm: Algorithms that leverage group theory and automorphism pruning are particularly effective for strongly regular graphs, where symmetry provides predictable patterns.

Graphs with High Symmetry

Highly symmetric graphs, such as complete graphs or regular polygons, have predictable structures that simplify isomorphism testing. For example, in a complete graph, every vertex connects to every other vertex, creating a highly regular structure.

Known Algorithm: Symmetry-finding algorithms that reduce the search space are used for graphs with high symmetry, often by collapsing equivalent vertices to streamline comparisons.

Bounded Treewidth Graphs

Graphs with bounded treewidth are graphs that can be decomposed into tree-like structures. Tree decomposition provides a way to represent the graph in a hierarchical form, which simplifies the isomorphism problem.

Known Algorithm: Algorithms that use dynamic programming on tree decompositions are effective for bounded treewidth graphs, allowing for polynomial-time isomorphism testing.

7.4 Applications of Graph Isomorphism in Chemistry and Compound Identification

Graph isomorphism has significant applications in chemistry, particularly in compound identification and structural analysis. By representing molecules as graphs, chemists can leverage graph isomorphism techniques to detect structural similarities, verify chemical databases, and identify compounds with equivalent structures. This approach is particularly valuable in molecular biology, pharmacology, and chemical informatics, where identifying molecular similarity or equivalence.

1. Representation of Chemical Compounds as Graphs

In chemistry, molecules can be represented as molecular graphs where:

Atoms correspond to vertices, often labeled by atom type (e.g., carbon, hydrogen, oxygen).

Bonds between atoms are represented as edges, which may be labeled based on bond type (e.g., single, double, or triple bonds).

For instance, the benzene molecule (C_6H_6) can be represented as a cyclic graph with six vertices (representing carbon atoms) connected by alternating single and double bonds. This graph-based representation of molecules allows for the application of graph theory concepts, including graph isomorphism, to analyze and compare chemical structures.

Graph isomorphism, in this context, allows chemists to determine whether two molecular structures are identical (isomorphic) or structurally different. In databases and chemical search engines, such isomorphism testing enables efficient identification of compounds with equivalent molecular structures, streamlining chemical research and drug discovery.

2. Compound Identification Using Graph Isomorphism

Chemical compound identification involves determining whether a given molecular structure matches an existing structure in a chemical database. Graph isomorphism testing plays a central role in this process by allowing chemists to compare a query molecule with stored structures to find identical or similar compounds.

Exact Structure Matching:

Exact matching is essential in cases where a chemist needs to confirm the identity of a compound. By comparing the molecular graph of a sample compound with graphs in a chemical database, isomorphism testing can quickly identify if an exact match exists.

Example: Suppose a chemist identifies an unknown sample suspected to be caffeine. By generating a molecular graph for the compound and running an isomorphism test against a chemical database, they can verify whether the sample's structure matches the known structure of caffeine.

Substructure Search:

Substructure searching identifies specific functional groups or structural motifs within a larger molecule. In pharmaceutical research, substructure searches enable chemists to locate compounds containing a specific active group (e.g., a benzene ring or a hydroxyl group), facilitating targeted drug design and compound screening.

Example: A researcher looking for compounds containing a phenol group can use subgraph isomorphism to scan a database for any molecular graph containing this substructure. This allows for rapid identification of compounds with the desired functional group, aiding in the selection of potential drug candidates.

3. Detecting Structural Similarities Between Compounds

Structural similarity between compounds is a key factor in fields like pharmacology, where similar structures may imply similar biological activity. Graph isomorphism and related techniques enable the detection of structural analogs, helping researchers identify compounds with potentially similar therapeutic effects.

Chemical Similarity Metrics:

By using graph isomorphism testing, chemists can compute similarity scores between molecular structures, providing a quantitative measure of how closely related two compounds are. Chemical similarity metrics help group compounds into families based on structural characteristics, which is particularly useful in chemical informatics and cheminformatics for clustering compounds with similar properties.

Example: In drug discovery, a similarity search might reveal that a newly developed compound have structure highly similar to that of a known therapeutic agent, suggesting potential efficacy and allowing researchers to prioritize further testing.

Case Study: Identifying Structural Analogues in Drug Design

In drug design, identifying structural analogs is crucial for optimizing molecular structures to enhance drug efficacy and reduce side effects. For instance, by using graph isomorphism to detect structural analogs of known inhibitors, researchers can design novel compounds with similar binding properties but improved therapeutic profiles. Pharmaceutical companies commonly use graph isomorphism tools to scan compound libraries for analogs of effective drugs, accelerating the development of new medications.

4. Verification and Management of Chemical Databases

Chemical databases store vast numbers of molecular structures, making efficient organization and verification essential. Graph isomorphism plays an important role in maintaining database integrity, ensuring each entry is unique, and verifying the structure of newly added compounds.

Structure Verification:

Graph isomorphism is used to confirm that a newly added compound structure does not already exist in the database under a different label or name. This is critical for maintaining accurate, non-redundant entries in chemical libraries, preventing duplication, and ensuring reliable data for researchers.

Example: If a chemist submits a new compound for addition to a database, an isomorphism check can confirm whether this compound is genuinely unique or simply a duplicate of an existing structure. This helps streamline database management and improve data accuracy.

Efficient Database Searches and Query Optimization:

Chemical databases, especially those used in drug discovery, contain millions of compounds, making search optimization essential. Graph isomorphism algorithms allow for efficient indexing and retrieval, enabling rapid queries based on structural similarity or exact matching.

Case Study in Database Management: The PubChem database, a widely used chemical database, uses graph-based algorithms to allow researchers to quickly search for compounds based on structure. By incorporating graph isomorphism, PubChem can retrieve compounds with similar structures, enabling faster drug discovery and molecular analysis.

5. Case Studies in Compound Identification and Structural Analysis

The use of graph isomorphism extends to real-world applications in chemical research, facilitating compound identification and structural analysis. Below are a few case studies that illustrate its impact.

Case Study 1: Fingerprint-Based Screening in Pharmaceutical Research

Fingerprint-based screening is a technique that encodes molecular structures as fingerprints, representing structural features such as atom types and bond connectivity. Graph isomorphism algorithms help generate these fingerprints, which can then be used to compare and retrieve compounds with similar structural features from large chemical libraries.

Application: In pharmaceutical research, fingerprint-based screening aids in the rapid identification of lead compounds by enabling structural comparison at scale. This approach has been instrumental in the early stages of drug discovery, where thousands of compounds are screened to identify candidates with specific therapeutic properties.

Case Study 2: Toxicology Analysis and Environmental Chemistry

In toxicology, identifying structurally similar compounds is crucial for predicting potential environmental hazards. For instance, polycyclic aromatic hydrocarbons (PAHs), a class of organic pollutants, are known to pose health risks. Graph isomorphism can help identify PAHs and related compounds in environmental samples, allowing chemists to assess contamination levels and prioritize cleanup efforts.

Application: Using molecular graphs to represent PAH compounds, researchers can perform similarity searches in environmental chemistry databases to locate related pollutants. This process aids in risk assessment and environmental monitoring by identifying pollutants with known toxicological profiles.

7.5 Current Challenges and Developments in Graph Isomorphism

The graph isomorphism problem is a foundational question in computer science and graph theory, yet it has long defied simple classification in computational complexity. While polynomial-time solutions exist for specific graph classes, the general case remains unsolved in the context of P versus NP. Recent breakthroughs, including Laszlo Babai's quasi-polynomial algorithm, have made significant strides, but challenges persist. This section examines the unresolved issues, recent advances.

1. Complexity Classification of Graph Isomorphism

The NP-Intermediate Status:

Graph isomorphism remains unique in its complexity classification, neither confirmed as P nor NP-complete. Traditionally classified as NP-intermediate, it does not fit comfortably in the existing complexity hierarchy. NP-intermediate problems were proposed by Richard Ladner, who hypothesized a category of NP problems that are neither solvable in polynomial time (P) nor reducible to NP-complete problems. Graph isomorphism is among the few problems with this classification.

Implications for Complexity Theory:

A polynomial-time solution would place graph isomorphism in P, confirming that it is computationally feasible. Conversely, a reduction from an NP-complete problem would shift it to NP-complete, confirming it as computationally intractable under typical assumptions. The ambiguous complexity class of graph isomorphism thus adds theoretical importance for problem, as it occupies a middle ground in the complexity hierarchy, making its solution a potential milestone in computational complexity theory.

2. Babai's Quasi-Polynomial Time Algorithm

In 2015, Laszlo Babai introduced a quasi-polynomial algorithm for graph isomorphism, marking a significant advancement in the field. Babai's algorithm, with a time complexity of $O(2^{(\log n)^c})$, narrows the gap between polynomial and exponential complexity, positioning graph isomorphism closer to P than to NP-complete problems.

Algorithmic Approach:

Babai's algorithm combines group theory, combinatorial techniques, and partition refinement to test isomorphism efficiently. The algorithm iteratively refines partitions of graph vertices and uses group-theoretic techniques to reduce the search space, thereby achieving quasi-polynomial time. Babai's use of combinatorial group theory was particularly innovative, allowing the algorithm to operate more efficiently on complex graph structures.

Implications of Babai's Algorithm:

Babai's work demonstrated that graph isomorphism is closer to polynomial time than previously thought, especially for dense and irregular graphs. While not a definitive solution, this quasi-polynomial breakthrough has inspired optimism for a possible polynomial-time algorithm and has spurred further research on algorithms that could yield polynomial complexity.

3. Ongoing Research Efforts and Optimizations in Isomorphism Testing

Research in graph isomorphism continues to focus on enhancing existing algorithms and developing solutions that are efficient across broader classes of graphs. Researchers are exploring approaches that combine graph automorphisms, advanced heuristics, and partitioning methods, while also refining algorithms tailored to specific graph types.

A. Automorphisms and Symmetry Exploitation

An automorphism is a mapping from a graph to itself that preserves its structure, effectively providing insight infor graph's symmetry. By identifying automorphisms, researchers can simplify isomorphism testing by reducing the number of required comparisons between vertices.

Symmetry-Based Pruning: Algorithms use graph symmetries to prune search spaces, avoiding redundant mappings in isomorphism testing. This symmetry-based pruning is particularly effective in highly regular graphs, where symmetrical patterns allow for a substantial reduction in computational steps.

Applications in Practical Graph Isomorphism Testing: Symmetry-exploiting algorithms are used in chemical compound databases and network analysis, where regular structures are common. Symmetry-aware approaches streamline isomorphism testing, enabling faster identification of equivalent graphs in large databases.

B. Parallel Processing and Quantum Computing

Parallel processing is another area of development in graph isomorphism, enabling the distribution of computations across multiple processors. This approach is particularly beneficial for large and complex graphs where exhaustive testing would otherwise be infeasible.

Parallel Isomorphism Algorithms: These algorithms divide the graph isomorphism task into smaller subproblems, each handled concurrently. While parallelization does not reduce the fundamental complexity, it speeds up the process, making isomorphism testing more feasible for large datasets.

Quantum Computing in Graph Isomorphism: Quantum algorithms, such as Grover's algorithm, have shown potential for accelerating isomorphism testing by leveraging quantum parallelism. Quantum approaches are in early stages, but research suggests that quantum computing could offer significant speedups, particularly for large-scale isomorphism problems.

C. Machine Learning and Heuristic Approaches

Machine learning techniques, especially in the field of graph neural networks (GNNs), are being applied to graph isomorphism testing. By learning representations of graph structures, machine learning models can classify graphs based on structural similarities, providing an alternative to traditional isomorphism testing methods.

Graph Neural Networks (GNNs): GNNs are capable of learning embeddings that capture the structural properties of graphs. These embeddings can then be used to approximate isomorphism, identifying graphs that are likely isomorphic without exhaustive testing. Although GNNs may not guarantee an exact solution, they offer a scalable approach for large datasets where approximate isomorphism is acceptable.

Limitations of Machine Learning Approaches: Machine learning methods may not provide guarantees of correctness, as ay are generally heuristic. Nonetheless, they are useful in scenarios where exact isomorphism is unnecessary, such as clustering similar molecules in chemical databases or categorizing social network structures.

4. Potential Future Directions

The ongoing research in graph isomorphism suggests several promising directions that may bring us closer to a definitive solution or at least improved methods for handling specific cases.

A. Further Refinement of Babai's Algorithm

Researchers continue to analyze and refine Babai's quasi-polynomial algorithm, exploring possibilities for optimization and simplification. Some researchers are working on reducing the complexity constants in Babai's approach or generalizing its techniques to additional graph classes. Efforts are also underway to adapt Babai's methods to handle larger graphs more efficiently, especially in fields like cheminformatics and bioinformatics, where dense and complex graphs are common.

B. Exploring Fixed-Parameter Tractability

Fixed-parameter tractable (FPT) approaches treat certain parameters as fixed, simplifying the problem under specific conditions. In graph isomorphism, researchers are examining ways to identify parameters that allow for FPT solutions. For example, FPT algorithms that treat treewidth, degree bounds, or other graph characteristics as fixed have shown potential in narrowing complexity for specialized cases.

Application in Specialized Graphs: By leveraging fixed-parameter tractability, researchers can develop efficient solutions for classes of graphs commonly found in practical applications, such as tree-like structures in network design or low-degree graphs in chemical structures.

C. Expansion of Symmetry-Aware and Dynamic Algorithms

Dynamic algorithms, which adjust calculations as new data arrives, are being explored as a solution for handling continuous updates in graphs, such as in network monitoring or real-time data processing. By integrating symmetry detection, dynamic algorithms could adapt more effectively to changes, preserving isomorphism properties without re-evaluating the entire graph.

Applications in Real-Time Systems: In real-time network analysis, where graphs evolve continuously, dynamic algorithms could monitor structural changes without complete re-evaluation. This approach holds potential for applications in cybersecurity, where detecting structural similarities in evolving networks is essential.

CHAPTER 8

BIPARTITE GRAPHS

8.1 Definition and Properties of Bipartite Graphs

In graph theory, bipartite graphs are a unique class of graphs with distinct structural properties that make them useful in various applications, from social networks to scheduling problems. Their defining feature is a division of vertices into two disjoint sets with specific adjacency constraints. This section provides a detailed introduction to bipartite graphs, including their formal definition, properties, and examples to highlight what differentiates bipartite graphs from other types of graphs.

1. Definition of Bipartite Graphs

A bipartite graph is a graph $G = (V, E)$ in which the vertex set V can be divided into two disjoint subsets V_1 and V_2 such that no two vertices within the same subset are adjacent. In other words, all edges in a bipartite graph connect a vertex from V_1 to a vertex from V_2, with no edges existing between vertices within V_1 or V_2 alone.

Formal Definition:

A graph $G = (V, E)$ is bipartite if its vertex set V can be partitioned into two subsets V_1 and V_2 such that for every edge $(u, v) \in E$, either $u \in V_1$ and $v \in V_2$, or $u \in V_2$ and $v \in V_1$. This requirement implies that no edge can exist between vertices within the same subset.

Example:

Consider a graph G with vertices $V = \{A, B, C, D\}$ and edges $E = \{(A, B), (A, D), (C, B), (C, D)\}$. We can divide V into two subsets: $V_1 = \{A, C\}$ and $V_2 = \{B, D\}$. Each edge connects a vertex in V_1 with a vertex in V_2, making G a bipartite graph.

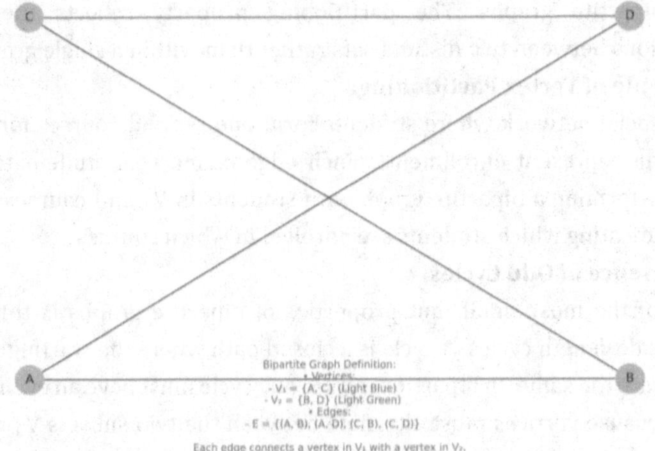

Figure 8.1 | Bipartite Graph

Definition:

A graph $G=(V,E)$ is bipartite if its vertex set V can be partitioned into two subsets V_1 and $V2 V_2 V2$ such that every edge connects a vertex in V1 with a vertex in V_2, with no edges within the same subset.

Graph Details:

- **Vertices:**
  - V1={A,C} (Light Blue)
  - V2={B,D}(Light Green)
- **Edges:**
  - E={(A,B),(A,D),(C,B),(C,D)}

8.2 Properties of Bipartite Graphs

Bipartite graphs have several unique properties that arise from their two-part structure. These properties include:

A. Vertex Partitioning

The defining characteristic of a bipartite graph is its vertex partitioning. This property ensures thfor vertex set V can be divided into two subsets V_1 and V_2, where edges only exist between vertices in different subsets. This partitioning gives bipartite graphs a predictable structure that distinguishes them from other types of graphs.

Implication for Real-World Applications: Many practical problems, such as job assignment and social network analysis, benefit from modeling relationships between two distinct groups (e.g., workers and tasks, people and affiliations) using bipartite graphs. The partitioning property reflects the need for connections between two distinct sets rather than within a single group.

Example of Vertex Partitioning:

In a social network where students form one set and courses form another, edges can represent enrollments. Each edge connects a student to a course, naturally forming a bipartite graph with students in V_1 and courses in V_2, with edges indicating which students are enrolled in which courses.

B. Absence of Odd Cycles

One of the most significant properties of bipartite graphs is thfory do not contain odd-length cycles. A cycle is a closed path where the starting and ending vertices are the same. In bipartite graphs, any cycle must have an even number of edges because vertices must alternate between the two subsets V_1 and V_2 as a cycle progresses.

Theorem: A graph is bipartite if and only if it contains no cycles of odd length.

Proof Outline: If a graph have odd cycle, it cannot be divided into two sets such that no two vertices in the same set are adjacent, since the odd cycle would force at least one vertex to connect back to a vertex within the same subset. Conversely, if a graph is bipartite, any cycle within it must alternate between V_1 and V_2, necessitating an even length.

Example of Absence of Odd Cycles:

Consider a triangle graph with three vertices A, B, and C connected cyclically by edges (A, B), (B, C), and (C, A). This cycle has three edges, an odd number, making it impossible to partition the vertices into two sets without an edge connecting vertices within the same set. Thus, this triangle graph is not bipartite.

In contrast, a square graph with vertices A, B, C, and D connected cyclically by edges (A, B), (B, C), (C, D), and (D, A) have cycle of four edges, an even number, making it a bipartite graph.

C. 2-Colorability

Bipartite graphs are 2-colorable, meaning that it is possible to assign one of two colors to each vertex so that no two adjacent vertices share the same color. The two colors represent the two subsets V_1 and V_2 of the vertex partition, with edges only connecting vertices of different colors.

Explanation of 2-Colorability:

Since bipartite graphs have no odd cycles, the vertices in one subset can be colored with one color, while the vertices in the other subset can be colored with a second color, ensuring that adjacent vertices always have different colors. This property can be a practical approach to quickly verifying if a graph is bipartite: if a graph is 2-colorable, it is bipartite.

Example of 2-Colorability:

Consider a graph with vertices V = {1, 2, 3, 4} and edges E = {(1, 2), (2, 3), (3, 4), (4, 1)}, forming a square. By coloring vertices 1 and 3 with one color (say, red) and vertices 2 and 4 with another color (say, blue), we achieve a 2-colorable graph with no two adjacent vertices sharing the same color, confirming thfor graph is bipartite.

3. Examples of Bipartite Graphs

Bipartite graphs are frequently encountered in real-world applications and theoretical studies. Below are some common examples:

Example 1: Matching Problems

In job assignment problems, bipartite graphs represent connections between two groups, such as workers and jobs. Each worker is represented by a vertex in V_1, each job by a vertex in V_2, and edges represent suitable matches between workers and jobs. Bipartite graphs ensure that workers only connect to jobs, never to other workers, naturally satisfying the bipartite structure.

Example 2: Social Networks

In social networks, bipartite graphs can model relationships between people and interests. One set of vertices represents users, and the other represents activities or groups, with edges indicating participation or interest. For instance, in an online platform, users might connect to groups based on interests, and this structure can be represented as a bipartite graph where users only connect to groups, not directly to each other.

Example 3: Biological Networks

In biological research, bipartite graphs are used to model relationships between two biological entities, such as species and habitats or genes and traits. For example, a bipartite graph can represent species in one set and habitats in another, with edges indicating which species live in which habitats. Similarly, in genomics, genes and traits might form a bipartite graph where edges show gene-trait associations.

8.3 Applications of Bipartite Graphs in Social and Job Networks

Bipartite graphs provide an effective way to model relationships between two distinct sets, which is particularly useful in social networks and job-matching contexts. By representing interactions between groups, bipartite graphs enable the analysis of connections and patterns that would be complex to manage with other graph structures. This section explores the applications of bipartite graphs in social and job networks, demonstrating how they facilitate problem-solving and data representation.

1. Bipartite Graphs in Social Networks

In social networks, bipartite graphs are often used to represent interactions between two sets of entities, such as individuals and interests or users and groups. This structure captures the relationships without introducing connections within a single set, which simplifies the modeling of diverse associations and enables effective data analysis.

A. Modeling Users and Interests

One of the most common applications of bipartite graphs in social networks is representing the relationship between users and interests. In this context:

Vertices in one subset V_1 represent users, while vertices in the other subset V_2 represent interests, such as hobbies, sports, or groups.

Edges represent connections between users and their interests, where each edge links a user to a specific interest.

This setup allows social network platforms to analyze and visualize the interests of their user base, identify popular topics, and recommend content based on shared interests.

Example:

Imagine a social network with users U = {Alice, Bob, Carol} and interests I = {Hiking, Photography, Gaming}. The relationships can be represented as a bipartite graph with the edges:

- ☐ (Alice, Hiking)
- ☐ (Bob, Photography)
- ☐ (Carol, Gaming)
- ☐ (Alice, Photography)

Using this bipartite representation, the network can easily track which users share similar interests and suggest relevant connections. If the network identifies that users with common interests are likely to engage more, they can leverage this data to make friend recommendations, enhancing user experience.

B. Group and Member Dynamics

Bipartite graphs are also useful in representing group memberships within social platforms. In this scenario, one subset of vertices represents individuals, and the other represents groups. Edges between them indicate membership, where each edge shows that a user belongs to a particular group.

Case Study: Facebook Groups

On Facebook, users join groups related to various topics, such as hobbies, professions, and local communities. A bipartite graph representation allows Facebook to identify communities with shared memberships and recommend groups to users based on their connections.

For example, if users interested in "Photography" are also likely to join groups on "Photo Editing," Facebook can recommend these groups to new users interested in photography, fostering stronger community engagement and better content discovery.

2. Bipartite Graphs in Job Matching

In job matching scenarios, bipartite graphs provide an efficient structure for representing potential assignments between job seekers and employers. Each edge in the graph represents a potential match, based on qualifications, preferences, or other criteria, facilitating the matching process.

A. Representing Job Seekers and Employers

In a job market context:

One subset V_1 represents job seekers, while the other V_2 represents employers or job positions.

Each edge signifies a possible match, where a job seeker is qualified or interested in a specific job posting.

This setup allows for straightforward modeling of potential assignments, enabling job platforms to analyze compatibility, recommend job opportunities, and facilitate optimal matches.

Example:

Consider a job-matching system with job seekers S = {John, Maria, Kevin} and job openings J = {Engineer, Designer, Analyst}. The possible matches are represented as edges:

- ☐ (John, Engineer)
- ☐ (Maria, Designer)
- ☐ (Kevin, Analyst)
- ☐ (Maria, Analyst)

By using this bipartite graph structure, the platform can analyze job compatibility based on qualifications and suggest the best matches for each job seeker. For instance, if Maria is suited for both Designer and Analyst positions, the platform might rank these matches based on her qualifications or preferences, guiding her application decisions.

B. Case Study: LinkedIn Job Recommendations

LinkedIn uses a similar bipartite structure to recommend job opportunities to its users. By representing job seekers and job postings as separate sets and matching users to relevant jobs based on their profiles, LinkedIn facilitates personalized job suggestions.

Implementation:

LinkedIn's recommendation algorithm identifies connections between user qualifications and job requirements. For instance, a user with experience in data science may be linked to job postings for "Data Analyst" or "Machine Learning Engineer" roles. This matching allows LinkedIn to suggest jobs that align with the user's skills, experience, and interests.

Bipartite graph modeling enables LinkedIn to efficiently search through its vast database of job postings, offering job seekers tailored recommendations and helping employers reach qualified candidates, enhancing the recruitment process.

3. Further Applications and Benefits

Bipartite graphs provide a flexible framework that can model many other types of relationships within social and job networks, helping address specific business needs and optimize the user experience.

A. Recommender Systems

Recommender systems are a core component of many platforms, and bipartite graphs are instrumental in creating recommendation algorithms that enhance user engagement. For example, streaming services like Netflix and Spotify use bipartite graphs to link users with content they might enjoy.

Vertices in V_1 represent users, while V_2 represents movies, songs, or other content.

Edges represent user engagement with the content (e.g., viewing, liking, listening), allowing the system to recommend similar items to users with shared interests.

By leveraging bipartite graphs, recommender systems can quickly identify patterns and similarities, suggesting relevant content to users and fostering increased engagement.

B. Educational Networks

In educational platforms, bipartite graphs can model relationships between students and resources, such as courses or projects. For instance:

Vertices in V_1 represent students, and those in V_2 represent available courses.

Edges signify course enrollment, allowing schools and online learning platforms to analyze course popularity and student preferences.

Platforms like Coursera and edX can use this structure to recommend courses based on student interests, track resource usage, and adjust their course offerings to better align with student demand.

C. Case Study: Internship Matching Platforms

Platforms focused on connecting students with internship opportunities, such as Handshake, use bipartite graphs to facilitate matches between students and employers offering internships. Each side of the bipartite graph consists of students and employers, with edges representing possible internship placements based on students' qualifications and employers' requirements.

Using this structure allows platforms like Handshake to quickly identify qualified candidates for specific internships, enabling employers to reach the most suitable students. Additionally, students benefit from a streamlined search for relevant opportunities, enhancing their chances of securing valuable internships that match their career goals.

8.4 Key Algorithms for Matching and Optimization in Bipartite Graphs

In bipartite graphs, efficient matching and optimization algorithms play a central role in solving pairing problems, such as job assignments, network matching, and resource allocation. Some of the most effective algorithms used for these purposes include the Hungarian algorithm, Hopcroft-Karp algorithm, and augmenting path methods. Each algorithm offers unique advantages in terms of complexity, applicability, and efficiency, particularly for bipartite graphs. This section provides an in-depth overview of e...

1. Hungarian Algorithm

The Hungarian algorithm, also known as a Kuhn-Munkres algorithm, is widely used for finding the maximum matching or minimum-cost matching in a weighted bipartite graph. This algorithm is particularly suited to solving the assignment problem, where the goal is to minimize the cost of assigning tasks to agents or resources.

How the Hungarian Algorithm Works:

1. Construct the Cost Matrix: Represent the bipartite graph as a matrix where rows represent agents (or one set of vertices) and columns represent tasks (or the other set of vertices). The elements in the matrix represent the cost of assigning an agent to a task.

2. Row and Column Reduction: Adjust the matrix by subtracting the minimum value in each row from all row elements, followed by subtracting the minimum value in each column from all column elements.

3. Covering Zeros with Minimum Lines: Cover all zeros in the matrix with a minimum number of horizontal and vertical lines. If the number of lines equals the matrix order, an optimal assignment is possible; otherwise, proceed for next step.

4. Adjustment of Uncovered Elements: Find the smallest uncovered element, subtract it from all uncovered elements, and add it to elements covered twice. Repefor zero-covering process until the required number of lines is achieved.

5. Construct the Optimal Assignment: Once the required number of lines is obtained, select zeros in the matrix to construct an assignment where each agent is matched with one task, minimizing the total cost.

Example of the Hungarian Algorithm:

Consider a cost matrix representing three workers and three tasks:

$$[4\ 1\ 3]$$
$$[2\ 0\ 5]$$
$$[3\ 2\ 2]$$

Applying the Hungarian algorithm will yield an optimal assignment with the minimum total cost.

Strengths and Applications:

The Hungarian algorithm provides an efficient solution for assignment problem with a time complexity of $O(n^3)$, making it suitable for smaller or moderately sized bipartite graphs.

It is effective in various applications, such as scheduling, resource allocation, and job assignments where costs or preferences exist.

2. Hopcroft-Karp Algorithm

The Hopcroft-Karp algorithm is designed to find the maximum matching in an unweighted bipartite graph. It operates by alternating between phases of searching for augmenting paths and updating the matching, achieving high efficiency for large bipartite graphs.

How the Hopcroft-Karp Algorithm Works:

1. **Initialize Matching:** Start with an empty matching M.

2. **Breadth-First Search (BFS) for Augmenting Paths:** Perform a BFS to find the shortest augmenting paths from unmatched vertices in one set to

unmatched vertices in the other. Augmenting paths are paths that alternate between edges not in M and edges in M, beginning and ending with unmatched vertices.

3. **Depth-First Search (DFS) for Path Augmentation:** For each vertex discovered in the BFS phase, use DFS to find disjoint augmenting paths that do not overlap with other paths.

4. **Update the Matching:** Each augmenting path found increases the size of the matching by one. Repefor BFS and DFS phases until no more augmenting paths can be found.

5. **Result:** Once all possible augmenting paths are found, the algorithm outputs the maximum matching.

Example of the Hopcroft-Karp Algorithm:

Consider a bipartite graph with two sets of vertices, V_1 = {A, B, C} and V_2 = {1, 2, 3}, and edges E = {(A, 1), (B, 2), (C, 3), (A, 2)}. Using the Hopcroft-Karp algorithm, we find the maximum matching by identifying augmenting paths and updating the matching iteratively.

Strengths and Applications:

The Hopcroft-Karp algorithm have time complexity of $O(\sqrt{n} * m)$, where n is the number of vertices and m is the number of edges, making it highly efficient for large bipartite graphs.

It is particularly effective in applications requiring maximum matching in unweighted bipartite graphs, such as pairing students to projects or assigning tasks to workers where no weighting is involved.

3. Augmenting Path Methods

The augmenting path method is a general approach to increasing the size of a matching in the Graph by identifying paths that alternate between edges in the matching and edges not in the matching. This technique underlies many matching algorithms, including the Hopcroft-Karp algorithm, and is used to iteratively improve the current matching.

How Augmenting Paths Work:

1. **Identify Free Vertices:** Begin with a partial matching M and identify unmatched vertices in both subsets of the bipartite graph.

2. **Find Alternating Paths:** Look for paths starting and ending with free vertices that alternate between edges in M and edges not in M.

3. **Augment Matching:** When an augmenting path is found, "flip" the edges along this path, adding edges not in M to M and removing edges that are already

in M. This increases the size of M by one.

4. **Repeat Until Optimal Matching:** Continue identifying augmenting paths until no more can be found, which indicates thfor current matching is maximum.

Example of Augmenting Path Method:

Consider a bipartite graph with vertices $V_1 = \{A, B\}$ and $V_2 = \{1, 2\}$, and edges $E = \{(A, 1), (B, 1), (B, 2)\}$. Starting with an empty matching, we find an augmenting path such as $A \rightarrow 1 \rightarrow B \rightarrow 2$, and update the matching accordingly. Repeating this process yields a maximum matching of size 2.

Strengths and Applications:

Augmenting path methods are versatile and can be applied to both weighted and unweighted bipartite graphs.

They are widely used in resource allocation, matching students to classes, or other settings where incremental improvement in matching is needed.

Comparative Overview of Algorithms

Algorithm	Purpose	Complexity	Best Use Case
Hungarian Algorithm	Minimum-cost matching	$O(n^3)$	Weighted bipartite matching, especially for assignments with costs
Hopcroft-Karp	Maximum matching	$O(\sqrt{n} * m)$	Large unweighted bipartite graphs
Augmenting Path	Incremental matching	Varies with implementation	General matching, versatile across various applications

Each of these algorithms serves distinct types of bipartite graph problems, with the Hungarian algorithm focusing on cost optimization, Hopcroft-Karp maximizing matches in unweighted graphs, and augmenting paths providing a flexible method for incremental matching improvements. Choosing the appropriate algorithm depends on factors like graph size, weighting requirements, and computational constraints.

8.5 Theoretical Aspects and Theorems in Bipartite Graphs

Bipartite graphs hold significant theoretical importance in graph theory due forir unique structural properties and applications in modeling relationships between two distinct sets. Several foundational theorems, including Hall's Marriage Theorem and Kőnig's Theorem, establish essential properties and offer valuable insights into matching and covering in bipartite graphs. This section explores these theorems, providing formal statements, explanations, and examples to demonstrate their implications and applications.

1. Hall's Marriage Theorem

Hall's Marriage Theorem is a fundamental theorem in graph theory that provides a criterion for a perfect matching in bipartite graphs. The theorem is commonly illustrated with the "marriage problem," where a set of individuals in one group seeks to be paired with members of another group based on certain compatibility requirements.

Formal Statement of Hall's Marriage Theorem:

Let $G = (V_1 \cup V_2, E)$ be a bipartite graph, where V_1 and V_2 represent two disjoint sets of vertices. A perfect matching exists in G from V_1 to V_2 if and only if for every subset $S \subseteq V_1$, the neighborhood $N(S) \subseteq V_2$ havet least as many vertices as S itself. Formally:

$|N(S)| \geq |S|$ for all subsets $S \subseteq V_1$.

This condition is known as a Hall condition. If it holds for all subsets of V_1, then a matching exists that pairs every vertex in V_1 with a unique vertex in V_2.

Explanation and Intuition:

The Hall condition essentially states that for a perfect matching to be possible, every subset of vertices in V_1 must be able to connect to an equal or larger number of vertices in V_2. If any subset S in V_1 has fewer neighbors in V_2 than its own size, then it is impossible to pair each vertex in S uniquely to a vertex in V_2, preventing a perfect matching.

Example of Hall's Marriage Theorem:

Consider two sets $V_1 = \{A, B, C\}$ and $V_2 = \{1, 2, 3\}$ with edges $E = \{(A, 1), (A, 2), (B, 2), (B, 3), (C, 1), (C, 3)\}$. To check if a perfect matching exists, we verify the Hall condition for each subset $S \subseteq V_1$:

- For $S = \{A\}$, $N(S) = \{1, 2\}$, and $|N(S)| = 2 \geq |S| = 1$.
- For $S = \{B\}$, $N(S) = \{2, 3\}$, and $|N(S)| = 2 \geq |S| = 1$.
- For $S = \{C\}$, $N(S) = \{1, 3\}$, and $|N(S)| = 2 \geq |S| = 1$.
- For $S = \{A, B\}$, $N(S) = \{1, 2, 3\}$, and $|N(S)| = 3 \geq |S| = 2$.

Since all subsets satisfy the Hall condition, a perfect matching exists in this bipartite graph. One possible perfect matching is $\{(A, 1), (B, 2), (C, 3)\}$.

Applications of Hall's Marriage Theorem:

Hall's Theorem is valuable in fields such as resource allocation, scheduling, and network design, where perfect pairings are needed between two distinct sets. It ensures that when the Hall condition is met, it is possible to achieve a complete pairing of resources or individuals, making it useful in task assignment and team formation.

2. Kőnig's Theorem

Kőnig's Theorem is a classical result that links maximum matching with minimum vertex covers in bipartite graphs. This theorem highlights a unique property of bipartite graphs by establishing an equivalence between the maximum size of a matching and the minimum size of a vertex cover.

Formal Statement of Kőnig's Theorem:

In any bipartite graph, the size of a maximum matching equals the size of a minimum vertex cover. If $G = (V_1 \cup V_2, E)$ is a bipartite graph, then:

maximum matching size = minimum vertex cover size.

Explanation and Intuition:

A vertex cover in the Graph is a set of vertices such that each edge in the graph is incident to at least one vertex in the cover. In bipartite graphs, Kőnig's Theorem guarantees thfor smallest number of vertices needed to cover all edges is equal for largest possible number of edges in a matching. This equality is unique to bipartite graphs and does not generally hold in non-bipartite graphs.

Example of Kőnig's Theorem:

Consider a bipartite graph with vertices $V_1 = \{A, B\}$ and $V_2 = \{1, 2\}$ and edges $E = \{(A, 1), (A, 2), (B, 1)\}$.

- A maximum matching in this graph is $M = \{(A, 2), (B, 1)\}$, with size 2.
- A minimum vertex cover is $C = \{A, 1\}$, which also has size 2.

Kőnig's Theorem holds as a size of the maximum matching (2) is equal for size of the minimum vertex cover (2).

Applications of Kőnig's Theorem:

Kőnig's Theorem havepplications in optimization problems where covering and matching are essential, such as in scheduling, network routing, and resource management. It provides a basis for efficient algorithms that find maximum matchings and minimum covers in bipartite graphs.

3. Other Significant Properties and Theorems in Bipartite Graphs

In addition to Hall's Marriage Theorem and Kőnig's Theorem, bipartite graphs exhibit several other important properties and theorems that distinguish them from other types of graphs.

A. Matching and Covering Relationships

Another important result in bipartite graphs is thfor matching number (the size of the largest matching) and the covering number (the size of the smallest vertex cover) are related uniquely due for graph's bipartite nature. This relationship helps simplify many combinatorial problems, making bipartite graphs highly useful in practical applications.

B. Maximum Flow and Minimum Cut

In bipartite graphs, the maximum flow-minimum cut theorem plays an integral role in network flow problems. This theorem states thfor maximum amount of "flow" from a source node to a sink node in a flow network is equal for minimum "cut" that separates the source from the sink. This concept is particularly relevant in bipartite graphs when solving problems such as transportation, network reliability, and connectivity in bipartite networks.

C. Bipartite Graph Properties

Bipartite graphs are characterized by the following structural properties, which have important theoretical implications:

2-Colorability: Bipartite graphs can be colored with two colors such that no two adjacent vertices share the same color. This property simplifies visualizations and can be used to solve scheduling conflicts.

Absence of Odd Cycles: A graph is bipartite if and only if it contains no cycles of odd length. This property aids in verifying whether a given graph is bipartite.

Examples and Applications of Theorems in Bipartite Graphs

The properties and theorems associated with bipartite graphs have significant applications in real-world scenarios, particularly in optimization and resource management.

Example Application of Hall's Marriage Theorem:

In a college course registration system, Hall's Theorem can be used to ensure that students are allocated to classes in the way that satisfies capacity constraints. If every subset of students has enough courses to meet their preferences, then a perfect matching is possible, ensuring that all students are assigned to courses they prefer.

Example Application of Kőnig's Theorem:

In an assembly line where tasks need to be assigned to workers with minimal redundancy, Kőnig's Theorem can be applied to determine the minimum number of workers required to cover all tasks. This approach optimizes task assignment, minimizing the workforce while covering all required tasks efficiently.

CHAPTER 9

GRAPH CYCLES

9.1 Types of Cycles in Graph Theory: Simple and Directed Cycles

In graph theory, cycles represent closed paths within the Graph, where a sequence of vertices and edges begins and ends for same vertex without traversing any edge more than once. Cycles can be found in both directed and undirected graphs, and they play a critical role in analyzing network structures, circuits, and various algorithms. This section delves infor two primary types of cycles-simple cycles and directed cycles-exploring their definitions, properties, and differences.

1. Simple Cycles in Undirected Graphs

Definition of Simple Cycle:

A simple cycle in an undirected graph is a closed path that begins and ends for same vertex and does not repeat any edge or vertex (except the starting and ending vertex). In simple cycles, each vertex and edge is visited only once within the cycle, ensuring no overlapping or retracing along the path.

Properties of Simple Cycles:

1. **No Repeated Edges or Vertices:** A simple cycle cannot contain repeated edges or vertices, aside from the starting and ending vertex.

2. **Closed Path:** A simple cycle is a loop within the graph, returning to its starting vertex.

3. **Minimum Number of Edges:** The simplest cycle have length of three edges, as a cycle must involve at least three vertices to return for starting point without reusing any vertex or edge.

Example of a Simple Cycle:

Consider an undirected graph G with vertices V = {A, B, C, D} and edges E = {(A, B), (B, C), (C, D), (D, A)}. A simple cycle in this graph could be:

$$A \rightarrow B \rightarrow C \rightarrow D \rightarrow A$$

This path satisfies the criteria of a simple cycle, as it starts and ends at vertex A without repeating any edges or vertices (other than the start/end vertex). If the graph contains additional connections, such as (A, C), other cycles might exist, but they too would require adherence for simple cycle properties.

2. Directed Cycles in Directed Graphs

Definition of Directed Cycle:

A directed cycle in a directed graph (digraph) is a sequence of directed edges that forms a closed loop, starting and ending for same vertex, where the direction of each edge follows a specific order. In a directed cycle, each edge (u, v) follows a directional path, meaning that if there's an edge from vertex u to vertex v, the reverse edge (v, u) cannot be part of the same cycle.

Properties of Directed Cycles:

1. **Directed Edges:** Each edge in a directed cycle have specific orientation, ensuring the cycle respects the directed nature of the graph.

2. **Ordered Path:** The sequence of vertices in a directed cycle follows the direction of edges without violating the graph's directional constraints.

3. **No Backtracking:** A directed cycle does not allow reverse traversal of any edge, making it distinct from undirected cycles where directionality is not a constraint.

Example of a Directed Cycle:

Consider a directed graph with vertices $V = \{A, B, C\}$ and edges $E = \{(A, B), (B, C), (C, A)\}$. A directed cycle in this graph is:

$A \rightarrow B \rightarrow C \rightarrow A$

This directed cycle starts and ends at vertex A, following the direction of each edge. If an additional edge (C, B) existed, it would not be part of this directed cycle, as each edge in the cycle follows a strict directional path.

Comparison with Simple Cycles:

While simple cycles in undirected graphs can include any path that loops back for start, directed cycles must strictly adhere to edge orientations. This property makes directed cycles more restrictive, as a sequence must follow the graph's specific directions.

Differences Between Simple and Directed Cycles

Property	Simple Cycles in Undirected Graphs	Directed Cycles in Directed Graphs
Edge Direction	No fixed direction for edges	Fixed direction for each edge
Vertex/Edge Repetition	No repeated edges or vertices (except start/ end vertex)	No repeated edges, follows directed path
Example Graph Structure	Road networks, utility networks	Traffic flow, computer networks
Complexity	More flexible due to lack of directional constraints	Requires specific directional order for each edge

Illustrative Examples

Example 1: Simple Cycle in a Social Network

In an undirected graph representing a social network, vertices represent individuals, and edges represent friendships. A simple cycle in this network could indicate a mutual set of friendships among a group, forming a closed friendship loop. For instance, in a network with vertices A, B, C, and D, if there are edges (A, B), (B, C), (C, D), and (D, A), the sequence:

$$A \rightarrow B \rightarrow C \rightarrow D \rightarrow A$$

represents a simple cycle, where each person in the loop is connected directly for next, with no retracing of friendships within the cycle.

Example 2: Directed Cycle in a Supply Chain Network

In a directed graph representing a supply chain, vertices might represent different companies, and edges represent supply relationships in a specific direction (e.g., supplier-to-buyer). A directed cycle could indicate a loop in the supply chain where each company in the cycle depends on the previous one in a circular manner. For example, if $A \rightarrow B \rightarrow C \rightarrow A$ forms a cycle in the supply chain, it implies that each company is involved in a reciprocal dependency, which can indicate potential inefficiencies or ...

9.2 Cycle Detection Algorithms in Graph Theory

Cycle detection is a fundamental task in graph theory with applications in fields like computer science, network analysis, and circuit design. Detecting cycles helps identify feedback loops, deadlocks, and potential redundancies within networks. Various algorithms are available for cycle detection in directed and undirected graphs, each tailored to different graph structures and use cases.

1. Depth-First Search (DFS) for Cycle Detection

The Depth-First Search (DFS) algorithm is one of the most intuitive and widely used methods for detecting cycles in both directed and undirected graphs. DFS is a graph traversal method that explores as far as possible along a branch before backtracking, making it well-suited for uncovering cycles.

9.3 Cycle Detection in Undirected Graphs Using DFS:

In undirected graphs, DFS detects cycles by marking each visited node and checking for back edges, which are edges that connect a node to an already-visited node other than its parent.

Algorithm Steps:

1. Start DFS from an arbitrary node and mark it as visited.
2. For each adjacent node, perform the following:

If the adjacent node is not visited, recursively call DFS on it, setting the current node as its parent.

If the adjacent node is visited and is not the parent of the current node, a cycle exists.

3. Continue the traversal until all nodes are visited.

Example in an Undirected Graph:

Consider a graph with vertices A, B, C, D and edges (A, B), (B, C), (C, D), (D, A), (B, D). Starting DFS from node A:

1. Visit A and mark it.
2. Move to B, mark it, and set A as a parent.
3. From B, move to C and mark it.
4. From C, move to D. Since D is already visited and is not the parent of C, a cycle A → B → C → D → A is detected.

Complexity: The DFS algorithm for cycle detection have time complexity of O(V + E), where V is the number of vertices and E is the number of edges, making it efficient for sparse graphs.

Cycle Detection in Directed Graphs Using DFS:

For directed graphs, DFS detects cycles by tracking recursion stack entries. If a vertex is revisited while it's still on the stack, a cycle exists.

Algorithm Steps:

1. Perform DFS on each unvisited vertex.
2. Mark each vertex as visited and add it for recursion stack.
3. For each adjacent vertex:

If it's unvisited, perform DFS recursively.

If it's already in the recursion stack, a cycle is detected.

4. Remove the vertex from the stack when backtracking.

Example in a Directed Graph:

Consider a directed graph with vertices A, B, C, D and edges (A, B), (B, C), (C, A), (C, D):

1. Starting DFS from A, add A for recursion stack.
2. Visit B, add it for stack.
3. Move to C, add it for stack.
4. From C, revisit A, which is in the recursion stack, indicating a cycle A → B → C → A.

2. Tarjan's Algorithm for Cycle Detection in Directed Graphs

Tarjan's Algorithm is a DFS-based method for finding strongly connected components (SCCs) in directed graphs, and it's highly effective for detecting cycles in directed graphs. If a component has more than one vertex or a self-loop, a cycle is present.

Algorithm Steps:

1.   Assign each vertex an index and a low-link value. The index represents the visit order, and the low-link value indicates the smallest reachable vertex index from that vertex.

2.   Perform DFS from each unvisited vertex, setting its index and low-link values.

3.   Add each visited vertex to a stack.

4.   For each adjacent vertex:

If it's unvisited, recursively apply DFS, updating the low-link value.

If it's in the stack, update the low-link of the current vertex.

5.   If the vertex's low-link equals its index, it forms an SCC. All vertices in the SCC are part of the cycle.

Example Using Tarjan's Algorithm:

Consider a directed graph with vertices A, B, C, D and edges (A, B), (B, C), (C, A), (C, D). Running Tarjan's algorithm:

1.   Start with A, assign it an index of 1 and a low-link of 1.

2.   Move to B, assign it an index of 2 and a low-link of 2.

3.   Continue to C, with index and low-link of 3.

4.   Revisit A, so C's low-link updates to 1.

5.   After DFS completes, A, B, and C form an SCC, indicating a cycle.

Complexity: Tarjan's algorithm runs in O(V + E) time and is efficient for dense directed graphs.

3. Union-Find Algorithm for Cycle Detection in Undirected Graphs

The Union-Find algorithm is a powerful method for cycle detection in undirected graphs, particularly effective for graphs represented as edge lists. This algorithm leverages two main operations-find and union-to check for cycles.

Algorithm Steps:

1.   Initialize each vertex as its own parent (self-rooted).

2.   For each edge:

Use the find operation to check the root of each vertex in the edge.

If both vertices have the same root, a cycle exists.

Otherwise, use the union operation to combine the two sets by updating the root of one vertex.

3. Repeat for all edges.

Example Using Union-Find:

Consider a graph with vertices A, B, C, D and edges (A, B), (B, C), (C, A), (C, D):

1. For (A, B), find shows different roots, so apply union.
2. For (B, C), find shows different roots, so apply union.
3. For (C, A), find shows the same root, indicating a cycle.

Complexity: The Union-Find algorithm runs in nearly constant time $O(\alpha(V))$, where α is the inverse Ackermann function, which is very small for practical inputs. It is efficient for dense graphs, especially with optimizations like path compression and union by rank.

Comparison of Cycle Detection Algorithms

Algorithm	Graph Type	Complexity	Best Use Case
DFS	Directed & Undirected	$O(V+E)O(V + E)O(V+E)$	Simple cycle detection in sparse graphs
Tarjan's Algorithm	Directed	$O(V+E)O(V + E)O(V+E)$	Detecting SCCs and cycles in dense directed graphs
Union-Find	Undirected	$O(\alpha(V))O(\alpha(V))O(\alpha(V))$	Cycle detection in undirected graphs with edge lists

9.4 Applications of Graph Cycles in Routing and Allocation

Graph cycles play a significant role in solving complex problems in routing and allocation, particularly within fields such as transportation, logistics, and resource allocation. By examining the role of cycles in these areas, we gain insights into optimizing routes, managing resources, and balancing loads in networked systems. This section explores how cycles in graphs contribute to efficient and practical solutions in these industries.

1. Transportation Networks and Cycle Applications

In transportation, graph cycles are essential for designing routes, managing flow, and preventing inefficiencies in circular transportation systems.

A. Circular and Return Routes: Cycles in transportation networks allow for the creation of circular or return routes, enabling transportation systems to provide continuous services. Urban bus networks often use cycles to connect major locations in a looped fashion, minimizing the need for turnarounds and reducing travel times. For instance, a bus route that forms a cycle around a city's key areas can provide seamless transportation for commuters, allowing them to board and alight at different points without the bus needing to retrace its path.

B. Round-Trip Route Optimization: For transportation systems requiring round-trips-like airline or railway schedules-cycles help in route optimization. These round-trip routes must cover multiple destinations and return for starting point efficiently. In air travel, cycles can be used to model multi-city flights where an airline offers a route that visits several cities before returning for origin, ensuring fuel efficiency and cost savings.

Real-World Example: Airline Route Planning Airlines like United Airlines and Southwest use cycles within their route networks to design layover schedules and circular routes. By using cycles, these airlines optimize fuel consumption and allocate planes to maintain constant service across high-demand locations, maximizing service efficiency.

2. Logistics and Delivery Systems

In logistics, cycles help improve delivery efficiency, optimize route planning, and support resource allocation.

A. Delivery Route Cycles: In delivery services, such as those operated by Amazon or FedEx, cycles enable efficient distribution by allowing delivery vehicles to return forir origin point after visiting multiple destinations. Delivery routes that follow cycles minimize fuel costs, reduce delivery times, and ensure that resources like fuel and manpower are used efficiently.

B. Load-Balanced Delivery Routes: Cycles can be used to create balanced routes in delivery systems. A balanced delivery route ensures thfor workload across delivery vehicles is distributed evenly. Cycles help in structuring routes where each vehicle travels a similar distance, preventing any single vehicle from becoming overloaded. For instance, a logistics company can divide a service area into cycles, assigning each cycle to a specific truck to balance the load and maximize delivery efficiency.

Real-World Example: FedEx Delivery Routes FedEx uses cycle-based routing to deliver packages efficiently, often structuring routes that loop back to distribution centers. This setup allows FedEx to efficiently cover delivery zones, allocate delivery trucks based on workload, and avoid unnecessary travel by creating optimal cycle paths in urban areas.

3. Resource Allocation and Load Balancing

Graph cycles are fundamental in systems where resources need to be allocated efficiently, as in load balancing across distributed networks or data centers.

A. Resource Allocation in Data Centers: Data centers and cloud service providers like AWS or Google Cloud utilize cycles to manage data traffic and server loads. By using cycles, they can route data in a loop to balance processing loads and prevent server overloading. Cyclic routing of data packets ensures that no single server or cluster becomes a bottleneck, which is essential for large-scale applications with heavy data traffic.

B. Load Balancing in Distributed Systems: Load balancing in distributed networks requires balancing the traffic across multiple nodes, ensuring no node is overwhelmed with requests. Cycles play a key role by offering paths that can redirect excess load to less-used nodes within a cycle, distributing traffic evenly across the network.

Real-World Example: Load Balancing at Google Data Centers Google uses cycles within its network infrastructure to balance server loads across data centers. Cycles help reroute traffic during high demand periods, using loops in network paths to avoid congestion, keep latency low, and ensure reliable data delivery to end-users worldwide.

4. Inventory Management and Supply Chains

In supply chains, cycles contribute to managing inventory flow, monitoring stock levels, and maintaining a continuous supply loop between manufacturers, suppliers, and retailers.

A. Inventory Cycles for Continuous Supply: In supply chain management, cycles allow for a steady, predictable flow of goods between suppliers, manufacturers, and retailers. A cycle-based supply chain structure ensures that goods are replenished at regular intervals, preventing shortages. By setting up inventory cycles, companies can plan stock deliveries more accurately and synchronize production schedules with supply needs.

B. Reverse Logistics: Cycles also support reverse logistics, where products are returned from consumers back to suppliers or manufacturers. Reverse logistics is often cyclical, as products or materials loop back for recycling, repair, or disposal. Companies with high product turnover, like electronics manufacturers, design supply chains with reverse cycles to ensure returned products are efficiently handled.

Real-World Example: Inventory Cycles in Automotive Supply Chains Automotive companies like Toyota use cycles to manage their just-in-time inventory systems. By setting up cyclical deliveries from parts suppliers to assembly plants, Toyota minimizes inventory holding costs, ensuring that components arrive precisely when needed in the production process.

5. Telecommunications and Network Systems

In telecommunications, graph cycles are essential for ensuring redundancy, maintaining reliable connections, and supporting load-balanced communication paths.

A. Redundant Network Paths: Cycles provide redundancy in network systems by establishing multiple paths between nodes. In a cycle-based network, if one connection fails, data can reroute through an alternate path, maintaining connectivity. This redundancy is vital for telecommunications networks, where continuous service is essential, and interruptions can lead to significant issues for customers.

B. Optimal Data Routing: Telecommunication networks, especially those handling high volumes of traffic like internet backbones, utilize cycles to balance data flow and reduce bottlenecks. Cycles ensure that data has multiple paths, enabling load balancing and efficient routing in networks. For example, ISPs often set up cyclical network paths to handle data surges, especially in metropolitan areas with dense connectivity requirements.

Real-World Example: Redundant Paths in Internet Backbone Networks Companies like AT&T and Verizon use cycles in their network infrastructure to provide redundancy. In the event of a network node failure, cycles allow for rerouting of data through alternative paths, maintaining service continuity and reducing downtime.

6. Energy Distribution and Utility Networks

In energy distribution, cycles ensure continuous power flow, manage load balancing, and support energy distribution across multiple locations.

A. Power Grid Cycles: In power grids, cycles provide redundancy and flexibility in power distribution. Electrical grids often use cyclic paths to ensure that if one path fails, electricity can be rerouted to maintain service. Cycles in power grids also allow for load balancing, where energy demand can be shifted across different paths to prevent overloading any single path.

B. Renewable Energy Distribution: Cycles are especially useful in renewable energy networks, where power from sources like wind and solar can vary. By setting up cyclic paths, energy distribution systems can balance renewable power with conventional power sources, ensuring a steady energy supply.

Real-World Example: Cycles in the European Power Grid The European electricity network uses cycles to manage power distribution across multiple countries. Cycles in this grid ensure cross-border electricity flow, allowing countries to share power reserves, stabilize supply during demand fluctuations, and support renewable energy integration.

9.5 Advanced Topics in Cycle Theory

Cycle theory is a cornerstone of graph theory, focusing on the study of cycles-closed paths in the Graph where the starting and ending vertices are the same. Advanced topics in cycle theory delve deeper infor structural, combinatorial, and algorithmic properties of cycles, exploring their roles in graph optimization, topology, and real-world applications.

Types of Cycles

Simple Cycle

A simple cycle is a closed path where no vertex or edge is repeated, except for the starting and ending vertex. Simple cycles form the basis for understanding more complex cyclic structures.

Hamiltonian Cycle

A Hamiltonian cycle is a special type of cycle that visits every vertex in the Graph for once and returns for starting vertex. Determining whether a graph contains a Hamiltonian cycle is NP-complete, making it a central problem in computational complexity.

Eulerian Cycle

An Eulerian cycle traverses every edge in the Graph for once and returns for starting vertex. A graph possesses an Eulerian cycle if and only if all vertices have even degrees, and the graph is connected.

Fundamental Cycles

Fundamental cycles arise from spanning trees. For any edge not included in the spanning tree, adding the edge for tree forms a cycle. These cycles are useful for analyzing the structure of graphs.

Directed Cycles

In directed graphs, cycles must follow the direction of edges. Directed cycles are central to topics like feedback systems, dependency resolution, and circuit design.

Cycles in Graph Optimization

Minimum Weight Cycle

In weighted graphs, finding the cycle with the minimum total weight is a common optimization problem. This is particularly important in applications like logistics, routing, and transportation.

Maximum Cycle Cover

The problem of finding a set of disjoint cycles that covers all vertices in the Graph, while maximizing the total weight, is another advanced area in optimization, often applied in scheduling and resource allocation.

Shortest Cycle Problem

The shortest cycle problem aims to identify the cycle with the least number of edges or smallest weight in weighted graphs. This problem havepplications in network design and failure detection.

Cycles in Graph Algorithms

Cycle Detection

Algorithms for detecting cycles are fundamental in various applications, such as deadlock detection in operating systems or feedback loop identification in circuits. Examples include:

Depth-First Search (DFS): Used for detecting cycles in undirected and directed graphs.

Union-Find Algorithm: Effective for detecting cycles in undirected graphs.

Cycle Enumeration

Enumerating all cycles in the Graph is a challenging computational task. Advanced algorithms like Johnson's algorithm enable efficient cycle enumeration in directed graphs.

3.3 Cycle Basis

A cycle basis is a minimal set of cycles from which all other cycles in the graph can be derived. Cycle bases are used in electrical circuit analysis and network reliability studies.

Topological Aspects of Cycles

Planarity and Cycles

Cycles play a critical role in determining the planarity of a graph. Kuratowski's theorem states that a graph is planar if it does not contain a subgraph homeomorphic to K_5 (complete graph on 5 vertices) or $K_{3,3}$ (complete bipartite graph).

Cycles in Surfaces

When studying graphs embedded on surfaces, cycles help in understanding the graph's genus (the number of "holes" in the surface). Non-contractible cycles (cycles that cannot be shrunk to a point) are of particular interest in topology.

CHAPTER 10

TREES IN GRAPH THEORY

10.1 Basic Concepts and Tree Terminology in Graph Theory

Trees are a foundational structure in graph theory, representing a unique class of graphs with specific properties that make them invaluable across data structures, algorithms, and various real-world applications. In this section, we'll introduce and explain the essential terms and structural characteristics that define trees, helping to establish a clear understanding of this important topic in graph theory.

1. Definition of a Tree

A tree is a type of connected, acyclic graph, which means it contains no cycles. In a tree, there is only one unique path between any two vertices, which makes it a highly structured and organized form of a graph. Formally, a tree $T = (V, E)$ is a graph where:

V is a set of vertices (or nodes), and

E is a set of edges connecting pairs of nodes.

For a graph to be a tree, it must be connected (all nodes must be reachable from any other node), and it must have no cycles.

Key Property: A tree with n nodes always has n - 1 edges. This is because any additional edge would create a cycle, violating the acyclic property of trees.

2. Essential Terminology in Trees

Understanding trees involves a series of terms that describe their structure and the relationships between their nodes. Below are some key terms and their meanings in the context of trees.

1. Node (or Vertex):

A node (or vertex) is a fundamental unit of a tree that can hold data or serve as a connector to other nodes in the structure.

Every tree is composed of nodes connected by edges, and each node represents a unique entity within the structure of the tree.

2. Edge:

An edge is a connection between two nodes in a tree. In a tree with n nodes, there are n - 1 edges.

Edges represent relationships between nodes, such as a link between a parent node and its child nodes.

3. Root:

The root is the topmost node in a tree, serving as a starting point or origin of the structure. In a rooted tree, all nodes are reachable from the root.

In hierarchical terms, the root is often considered the "ancestor" of all other nodes in the tree, and it has no parent.

4. Parent-Child Relationship:

A parent node is one that have direct edge to another node, known as its child. Each child node is directly connected to only one parent node in a tree.

For example, in a family tree, a parent node might represent a person with children as air descendants, directly connected below them.

5. Leaf Node (or Leaf):

A leaf node (or simply leaf) is a node in a tree that has no children; it is an endpoint in the structure of the tree.

Leaf nodes represent the "end" of paths within the tree and are often associated with data storage or final outcomes in certain applications, like decision trees.

6. Internal Node:

An internal node is any node in the tree that is not a leaf or the root. Internal nodes have at least one child.

These nodes often represent intermediate steps or decision points in tree-based processes or algorithms.

7. Degree:

The degree of a node is the number of children it has in the tree. A node with degree zero is a leaf node.

The degree of the root is also important, as it often influences the overall structure and branching of the tree.

8. Depth of a Node:

The depth of a node is the number of edges from the root node to that particular node. For instance, the root have depth of 0, and its children have a depth of 1.

Depth helps in determining the level at which a node exists within the tree structure.

9. Height of a Tree:

The height of a tree is the longest path from the root node to any leaf node, measured in edges. Alternatively, it can be thought of as a maximum depth of any node in the tree.

The height is crucial for understanding the efficiency of certain tree-based operations, especially in balanced trees where height is minimized.

10. Subtree:

A subtree is any section of a tree that forms a tree on its own. For any given node, the subtree rooted at that node includes that node and all of its descendants.

Subtrees are useful in recursive algorithms, where operations are applied to individual subtrees within a larger tree.

3. Structural Properties of Trees

Trees exhibit several structural properties that distinguish them from other types of graphs. These properties are useful for characterizing trees and understanding their applications.

1. Acyclic Nature:

A tree contains no cycles, meaning there is only one path between any two nodes. This property ensures that trees are highly organized structures with minimal redundancy.

2. Connectedness:

Trees are connected graphs, meaning that every node is reachable from the root. If a node were unreachable, the structure would be a forest (a collection of disjoint trees) rather than a single tree.

3. Uniqueness of Paths:

In a tree, there is exactly one path between any two nodes. This property enables efficient traversal and search operations, as are are no duplicate paths to check.

4. Recursive Structure:

Trees can be recursively defined, as each node (along with its descendants) forms a subtree. This recursive structure makes trees well-suited for recursive algorithms and data organization.

5. Hierarchy Representation:

Trees naturally represent hierarchies, making them ideal for modeling structures such as organizational charts, file systems, and taxonomies, where elements are nested within broader categories.

4. Examples Illustrating Basic Tree Terminology

Example 1: Family Tree Structure

A family tree is a common example of a tree where each person is a node, and edges represent parental relationships. The root might be an ancestor, with parent-child edges connecting descendants. For instance:

The root node represents the most distant ancestor.

Each parent has edges forir children, creating a hierarchical family structure.

Leaf nodes represent individuals with no descendants.

Example 2: Directory Structure in Operating Systems

File systems in operating systems are often represented as trees:

The root node represents the main directory (e.g., "C:" in Windows).

Each directory has subdirectories (child nodes), and files within each directory are leaf nodes with no further children.

This structure allows efficient organization and retrieval of files, with each file or folder accessible through a unique path from the root.

Example 3: Binary Tree Structure

A binary tree is a specific type of tree where each node have maximum of two children (often referred to as a left and right child).

Binary trees are used in data structures like binary search trees and heap data structures.

The structure of binary trees allows efficient searching, insertion, and deletion operations, particularly in balanced forms where the height is minimized.

10.2 Spanning Trees and Algorithms in Graph Theory

A spanning tree is a crucial concept in graph theory, especially for connected, undirected graphs. Spanning trees form the foundation for many applications, such as optimizing network designs and finding efficient routes. This section provides an overview of spanning trees, explains their properties, and explores the two primary algorithms used to find them: Kruskal's Algorithm and Prim's Algorithm.

1. Definition and Properties of Spanning Trees

A spanning tree of a connected, undirected graph $G = (V, E)$ is a subgraph that includes all vertices of G, is acyclic, and is connected. In essence, a spanning tree is a tree that spans all vertices in the Graph without forming any cycles, connecting them with the minimum number of edges.

Key Properties of Spanning Trees:

1.   **Number of Edges:** A spanning tree with n vertices has exactly n - 1 edges.

2.   **Acyclic and Connected:** Spanning trees are acyclic by definition and connect all vertices, meaning there is a unique path between any pair of vertices in the spanning tree.

3. Minimal Edge Set: Removing any edge from a spanning tree would disconnect it, while adding an edge would create a cycle, which reaffirms its minimal edge structure.

Example:

Consider a graph G with vertices V = {A, B, C, D} and edges E = {(A, B), (A, C), (B, C), (B, D), (C, D)}. One possible spanning tree for G could include the edges (A, B), (B, D), (C, A), which connects all vertices without forming a cycle and includes exactly three edges.

2. Minimum Spanning Trees (MST)

In a weighted graph, each edge have associated weight or cost, often representing distance, cost, or other metrics. A Minimum Spanning Tree (MST) is a spanning tree with the smallest possible sum of edge weights, making it essential for optimizing routes and minimizing costs in network designs.

Applications of MSTs:

1. Network Design: MSTs are used in designing efficient communication and transportation networks, ensuring all points are connected with minimal infrastructure costs.

2. Clustering: MSTs can help cluster data points by connecting them with minimum distances, useful in data analysis and machine learning.

3. Kruskal's Algorithm for Finding MSTs

Kruskal's Algorithm is a greedy algorithm that finds the MST by adding edges in increasing order of weight while avoiding cycles. This approach is especially efficient in sparse graphs.

Steps of Kruskal's Algorithm:

1. Sort Edges: Begin by sorting all edges in non-decreasing order of weight.

2. Initialize Sets for Each Vertex: Treat each vertex as a separate set.

3. Select Edges: Starting from the lowest-weight edge, add edges for MST if they connect two different sets (i.e., if they do not form a cycle). Use the Union-Find data structure to keep track of the connected components.

4. Repeat Until n-1 Edges are Included: Continue this process until the MST includes n - 1 edges.

Complexity of Kruskal's Algorithm:

Time Complexity: O(E log E) due to sorting the edges, where E is the number of edges.

Space Complexity: O(V), where V is the number of vertices, due to storing parent and rank information in the Union-Find structure.

Example Using Kruskal's Algorithm:

Consider a graph with vertices V = {A, B, C, D} and weighted edges:

- (A, B) = 1
- (A, C) = 3
- (B, C) = 2
- (B, D) = 4
- (C, D) = 5

Steps:

1.   Sort edges: (A, B), (B, C), (A, C), (B, D), (C, D).
2.   Add edge (A, B) for MST.
3.   Add edge (B, C), forming no cycle.
4.   Add edge (A, C), which would form a cycle, so skip it.
5.   Add edge (B, D), completing the MST with edges (A, B), (B, C), (B, D).

This MST have total weight of 1 + 2 + 4 = 7.

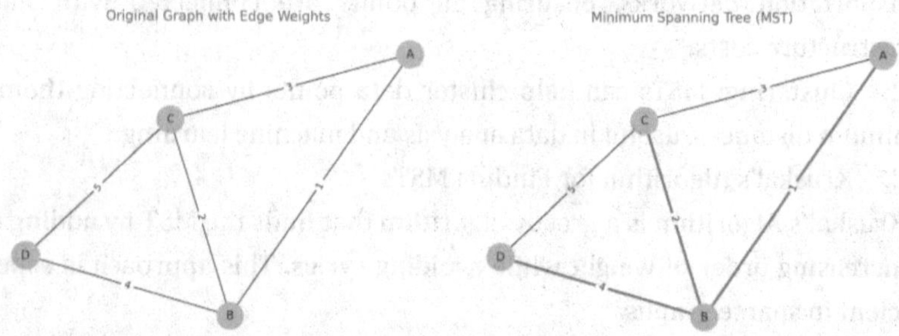

Figure 10.1 | Kruskal's Algorithm: Original Graph and MST

Left: Original Graph with Edge Weights

- **Vertices:** V={A,B,C,D}
- **Edges with Weights:**
  o   (A,B)=1
  o   (B,C)=2
  o   (A,C)=3
  o   (B,D)=4
  o   (C,D)=5

Right: Minimum Spanning Tree (MST)

- **Edges in MST:** (A,B),(B,C),(B,D)
- **Total Weight:** 1+2+4=7

- **Highlighted Edges:** Form the MST with no cycles.

4. Prim's Algorithm for Finding MSTs

Prim's Algorithm is another greedy algorithm that builds the MST by expanding from a starting vertex, adding edges that connect the growing MST to new vertices. Prim's is particularly efficient for dense graphs.

Steps of Prim's Algorithm:

1. Initialize MST: Start from an arbitrary vertex, adding it for MST.

2. Select Minimum-Weight Edge: Among the edges connected for current MST, choose the edge with the smallest weight that connects to a vertex not yet in the MST.

3. Expand MST: Add the chosen edge and the connected vertex for MST.

4. Repeat Until All Vertices are Included: Continue selecting minimum-weight edges that connect new vertices until the MST spans all vertices.

Complexity of Prim's Algorithm:

Time Complexity: $O(E \log V)$ with a priority queue, where E is the number of edges and V is the number of vertices.

Space Complexity: $O(V)$ for storing vertex data in the priority queue.

Example Using Prim's Algorithm:

Using the same graph as above:

1. Start from vertex A.

2. Choose edge (A, B) with weight 1, connecting B for MST.

3. From A and B, the smallest edge is (B, C) with weight 2.

4. Next, the smallest edge is (B, D) with weight 4, connecting D for MST.

The MST for this graph includes edges (A, B), (B, C), (B, D) with a total weight of 7.

5. Comparison of Kruskal's and Prim's Algorithms

Aspect	Kruskal's Algorithm	Prim's Algorithm
Approach	Sort edges by weight, then add while avoiding cycles	Expand MST from a starting vertex
Best for Graph Type	Sparse graphs	Dense graphs
Complexity	$O(E \log E)$	$O(E \log V)$ with a priority queue
Data Structures	Union-Find	Priority queue (often implemented with heaps)

10.3 Binary Trees and Their Properties in Graph Theory

Binary trees are a fundamental type of tree structure in graph theory and data structures, characterized by their unique branching pattern in which each node has, at most, two children. This section explores the structure, properties, and traversal methods of binary trees, with a focus on various types of binary trees and their specific characteristics.

1. Definition of a Binary Tree

A binary tree is a tree data structure in which each node havet most two children. These children are typically referred to as a left child and right child, creating a clear hierarchical structure that lends itself well to searching, sorting, and hierarchical data storage.

Properties of Binary Trees:

1. **Binary Structure:** Each node has up to two children, creating a branching factor of two at each level.

2. **Unique Path to Root:** Each node have single path for root of the tree, and there is only one path between any pair of nodes.

2. Types of Binary Trees

Binary trees come in various forms, each with unique properties that make them suited to specific applications and operations. Below are some primary types of binary trees:

A. Full Binary Tree:

In a full binary tree, every node has either 0 or 2 children, with no nodes having only one child. This structure creates a balanced and regular form where nodes at each level either have two children or none at all (i.e., they are leaf nodes).

Example: Consider a tree with nodes arranged as follows:

Root has two children, each child has two children of its own, and all leaves are for same level.

Full binary trees are often used in situations where data must be organized uniformly, as in decision trees for some classification problems.

B. Complete Binary Tree:

A complete binary tree is a binary tree where all levels are fully filled, except possibly the last level, which is filled from left to right.

This arrangement ensures thfor tree remains as compact as possible, optimizing space and structure.

Example: In a tree with three levels, the first two levels are completely filled, and nodes in the third level are filled from left to right.

Complete binary trees are commonly used in heap structures and priority queues, as air arrangement allows efficient access for minimum or maximum element.

C. Balanced Binary Tree:

A balanced binary tree is a binary tree in which the height difference between the left and right subtrees of any node is minimal, typically constrained by specific balancing criteria.

Balanced trees, like AVL trees and Red-Black trees, are widely used in databases and file systems where efficient data retrieval is necessary.

Balanced binary trees have logarithmic height relative for number of nodes, which optimizes search, insert, and delete operations.

D. Perfect Binary Tree:

A perfect binary tree is a complete binary tree in which all internal nodes have two children, and all leaf nodes are for same level.

Perfect binary trees are also called fully balanced binary trees and have an exact number of nodes, making them ideal for theoretical models and symmetrical data representations.

3. Properties of Binary Trees

Binary trees have properties that dictate their height, depth, and capacity. These properties help in understanding the efficiency of binary tree operations and their applications.

Height of a Binary Tree:

The height of a binary tree is the length of the longest path from the root to a leaf. In balanced binary trees, the height is minimized to $O(\log n)$, where n is the number of nodes.

The height directly affects the efficiency of operations like searching, insertion, and deletion, as ase are typically proportional for height of the tree.

Depth of a Node:

The depth of a node is the number of edges from the root to that node. Depth is crucial in determining the level of a node within the tree hierarchy.

The depth helps calculate traversal efficiency, as nodes closer for root are accessed more quickly.

Number of Nodes:

In a perfect binary tree of height h, the number of nodes n is given by:

$$n = 2^{(h+1)} - 1$$

This relationship allows for determining the maximum capacity of a binary tree given its height, or conversely, the minimum height required for a given number of nodes.

4. Traversal Methods in Binary Trees

Traversal in binary trees involves visiting each node in a systematic order, with three primary traversal methods: preorder, inorder, and postorder. Each method serves different purposes and yields different node sequences.

A. Preorder Traversal (Root, Left, Right):

In preorder traversal, the root node is visited first, followed by the left subtree, then the right subtree.

Order of Visit: Root → Left → Right.

Example Sequence: For a tree with root A, left child B, and right child C, the preorder sequence is A, B, C.

Preorder traversal is used in tasks like copying a tree structure or evaluating prefix expressions in expression trees.

B. Inorder Traversal (Left, Root, Right):

In inorder traversal, the left subtree is visited first, followed by the root, and then the right subtree.

Order of Visit: Left → Root → Right.

Example Sequence: For a tree with root A, left child B, and right child C, the inorder sequence is B, A, C.

Inorder traversal is widely used in binary search trees (BSTs) because it visits nodes in ascending order, making it suitable for sorted data retrieval.

C. Postorder Traversal (Left, Right, Root):

In postorder traversal, the left subtree is visited first, followed by the right subtree, and finally the root.

Order of Visit: Left → Right → Root.

Example Sequence: For a tree with root A, left child B, and right child C, the postorder sequence is B, C, A.

Postorder traversal is used in deleting nodes or evaluating postfix expressions, as it processes child nodes before their parent nodes.

5. Examples of Binary Trees

Example 1: Full Binary Tree

Consider a full binary tree with nodes A, B, C, D, E:

```
   A
  / \
  B  C
```

```
    /\
   D E
```

In this tree:

Node A has two children, and nodes B and C also have either two or zero children, meeting the full binary tree criteria.

Example 2: Complete Binary Tree

Consider a complete binary tree with nodes A, B, C, D, E:

```
    A
   /\
  B C
  /\
 D E
```

In this tree:

The first two levels are fully filled, and the last level is filled from left to right, matching the complete binary tree requirements.

Example 3: Balanced Binary Tree

An example of a balanced binary tree could look like this:

```
     F
    / \
   D   J
  /\   /\
 B E  G K
```

In this tree:

ach subtree havepproximately the same height, optimizing the tree for efficient operations.

10.4 Applications of Trees in Data Structures and System Design

Trees are a versatile data structure widely used in data management and system design due forir hierarchical and organized structure. Their application spans multiple domains, including file systems, databases, network routing, and organizational hierarchies. By facilitating efficient storage, retrieval, and management of data, trees are indispensable in both computational and real-world systems. This section explores key applications of trees in various domains, illustrating how they enhance function...

1. File Systems and Directory Trees

File systems commonly employ tree structures to organize directories and files. The hierarchical nature of trees provides an ideal framework for representing files and folders, where each directory is treated as a node that can contain other directories (subdirectories) or files (leaf nodes). The root directory represents the starting point of the hierarchy, and each subdirectory can have its own set of nested folders and files.

Example: Directory Tree in Operating Systems

In many operating systems, the file system is visualized as a tree structure. For instance, in Windows, the directory structure starts from the root (e.g., "C:\") and branches into folders like "Program Files," "Users," and "Windows." Each of these folders may further branch into additional subfolders and files.

Advantages: Directory trees allow efficient navigation and searching. By traversing the tree, users or applications can quickly locate specific files without scanning the entire file system.

Usage: Most operating systems use trees for file management due forir organized and recursive structure, which simplifies access, permissions, and organization.

2. Databases and Binary Search Trees (BST)

In databases, tree structures are essential for organizing data in the way that allows for efficient searching, inserting, and deleting of records. Binary Search Trees (BSTs) are widely used in databases for indexing and managing sorted data, as ay maintain a structured order that enables quick access to data.

Binary Search Trees (BST) in Databases

A Binary Search Tree (BST) is a binary tree where each node havet most two children, and each node's left child contains a value smaller than the node, while the right child contains a value larger than the node.

Advantages: BSTs enable fast lookups, insertions, and deletions, with average time complexity $O(\log n)$ for balanced trees. In databases, where data retrieval speed is crucial, this performance gain is significant.

Example: In a database managing student records sorted by ID, a BST allows efficient searching by ID number. Starting for root node, each comparison narrows down the possible location of the desired ID by half.

Usage: BSTs are used in indexing database rows, allowing databases to support efficient query responses. Balanced trees like AVL or Red-Black trees are often implemented to ensure thfor tree remains balanced and performant.

3. Network Routing and Spanning Trees

Tree structures play a fundamental role in network routing and topology design. In networking, spanning trees are often used to manage and organize the network structure, ensuring efficient data flow while preventing loops, which can cause data congestion.

Spanning Tree Protocol (STP) in Networks

Spanning Tree Protocol (STP) is a protocol used in Ethernet networks to create a loop-free logical topology. STP organizes the network into a spanning tree, where there is only one active path between any two nodes, preventing data loops.

Advantages: By implementing STP, network administrators ensure that redundant links are managed without causing network loops, which improves network reliability and efficiency.

Example: In a network with multiple switches, STP will deactivate redundant paths, keeping only the optimal path active. If a primary path fails, STP can automatically activate a redundant path.

Usage: Spanning trees are critical in large networks where multiple redundant paths exist, such as data centers and enterprise networks. This ensures reliability and reduces the risk of broadcast storms.

4. Organizational Hierarchies and Decision Trees

Trees are well-suited for representing hierarchical structures, making them ideal for modeling organizational hierarchies and decision processes. In an organizational context, a tree structure can represent reporting lines and departmental hierarchies, while in decision-making, trees are useful for laying out choices and their potential outcomes.

Organizational Hierarchy Trees

In a company hierarchy, the CEO may represent the root node, with direct reports branching out as child nodes. Each level of management corresponds to a level in the tree, and each node represents a specific position or department.

Advantages: Tree structures in organizations clarify reporting relationships and simplify hierarchical navigation within the organization.

Example: In a multinational corporation, a tree can represent various branches, each with sub-branches for regional managers, department heads, and individual employees.

Usage: Organizational trees are commonly used in Human Resource Management Systems (HRMS) and project management tools to structure teams and projects effectively.

Decision Trees in Data Science and Machine Learning

A decision tree is a tree-like model used for decision-making and classification tasks in data science. Each internal node represents a decision point or test on an attribute, and each leaf node represents a class label or outcome.

Advantages: Decision trees are interpretable and can handle both numerical and categorical data, making them popular in classification and regression tasks.

Example: In a loan approval model, a decision tree might include decision nodes based on factors like credit score, income, and loan amount, ultimately classifying an applicant as "approved" or "rejected."

Usage: Decision trees are widely used in data analytics, predictive modeling, and machine learning, providing insights into data patterns and aiding in decision-making.

5. Expression Trees in Compilers and Parsing

Expression trees are used in compilers and parsers to evaluate mathematical expressions and manage syntax structures. An expression tree represents an arithmetic or logical expression in tree form, where internal nodes are operators, and leaf nodes are operands.

Application of Expression Trees in Compilers

In an expression tree, each internal node represents an operator (e.g., +, -, *, /), while each leaf node represents an operand (e.g., variables or constants). This structure allows for systematic evaluation of expressions.

Advantages: Expression trees facilitate the evaluation of complex expressions by following a recursive pattern, evaluating from the lowest-level nodes upwards.

Example: For the expression (3 + 5) * 2, the corresponding expression tree would have "*" as a root, with "+" and "2" as children, and "3" and "5" as children of the "+" node.

Usage: Expression trees are widely used in parsing expressions in programming languages and in managing computations in calculators and interpreters.

6. Search and Compression Algorithms

Tree structures are essential in various algorithms, particularly in search and data compression. For instance, Trie and Huffman trees are specialized trees used in these contexts.

Trie Trees in Search Operations

A Trie is a tree-based data structure used for storing a dynamic set of strings, allowing fast search operations, particularly useful in text search engines.

Advantages: Trie trees enable efficient storage and retrieval of strings, with search operations performed in O(L), where L is the length of the string.

Example: In an autocomplete system, a Trie can store a dictionary of words, allowing rapid retrieval of suggestions based on prefix matching.

Usage: Tries are used in applications requiring fast prefix searches, like search engines, spell checkers, and autocomplete functions.

Huffman Trees in Data Compression

A Huffman Tree is a binary tree used in data compression algorithms. It represents characters based on their frequency of occurrence, assigning shorter codes to frequent characters and longer codes to rare characters.

Advantages: Huffman encoding reduces the amount of space needed to store data, making it suitable for file compression.

Example: In compressing a text document, frequent letters like "e" and "a" may receive shorter codes, while less common letters like "z" may receive longer codes.

Usage: Huffman trees are used in file compression algorithms, such as JPEG and MP3, to reduce file sizes while preserving data integrity.

10.5 Advanced Topics in Tree Theory

Tree theory is a fundamental branch of graph theory that studies trees-connected, acyclic graphs. Trees serve as a versatile tool in diverse applications, from computer science and biology to network design and optimization. Advanced topics in tree theory delve deeper inforir structural, combinatorial, and algorithmic properties, exploring the powerful insights they provide in both theoretical and practical contexts.

Types of Trees

Spanning Trees

A spanning tree of a graph is a subgraph that connects all vertices of the original graph without forming any cycles. Every connected graph havet least one spanning tree. Key properties include:

Minimum Spanning Tree (MST): A spanning tree with the smallest possible total edge weight. Algorithms such as Kruskal's and Prim's are widely used to find MSTs.

Maximum Spanning Tree: A spanning tree with the largest possible total edge weight, useful in applications like clustering.

Binary Trees

Binary trees are trees where each vertex havet most two children. They are central in computer science applications, particularly in search and sorting algorithms, and can be further classified into:

Full Binary Trees: Each vertex has 0 or 2 children.

Complete Binary Trees: All levels except the last are fully filled, and the last level havell nodes as far left as possible.

Balanced Binary Trees: Heights of the left and right subtrees differ by at most one at every node, ensuring efficiency in operations.

Rooted Trees

In rooted trees, one vertex is designated as a root, and all edges are directed away from or toward it. Rooted trees are used in hierarchical data structures and representations like file systems or organizational charts.

Steiner Trees

Steiner trees are used in network design, where the goal is to connect a subset of vertices (called terminals) with the minimum total edge weight. This problem generalizes the spanning tree problem and is NP-hard.

Decision Trees

Decision trees are rooted trees used in machine learning for classification and regression tasks. They provide an intuitive way to model decision-making processes.

Structural Properties of Trees

Diameter and Radius

Diameter: The longest path between any two vertices in a tree. It is used in network analysis to evaluate the efficiency of communication.

Radius: The minimum eccentricity (maximum distance from a node to any other node) of all vertices in the tree.

Tree Decomposition

Tree decomposition breaks a graph into a tree-like structure, where each node of the tree represents a subset of vertices of the graph. It is particularly useful in algorithms for solving NP-hard problems on graphs with bounded treewidth.

Treewidth

Treewidth measures how close a graph is to being a tree. Graphs with small treewidth allow for efficient solutions to problems like vertex coloring and constraint satisfaction using dynamic programming.

Algorithms in Tree Theory

Tree Traversal Algorithms

Traversal algorithms are fundamental in tree theory, with various techniques tailored for specific applications:

Depth-First Search (DFS): Explores as far as possible along each branch before backtracking. DFS is used in cycle detection, topological sorting, and game search trees.

Breadth-First Search (BFS): Explores neighbors level by level, ideal for finding the shortest path in unweighted graphs.

Inorder, Preorder, and Postorder Traversals: Specific to rooted binary trees, these traversals have applications in expression tree evaluations, XML parsing, and more.

Tree Center and Centroid

Tree Center: Vertices that minimize the maximum distance to other vertices in the tree. Centers are crucial for designing efficient networks and communication systems.

Tree Centroid: Vertices that minimize the sum of distances to all other vertices. Centroids are used in hierarchical clustering and data analysis.

Tree Pruning

Tree pruning involves removing unnecessary nodes or branches to optimize performance or simplify structure. For example, alpha-beta pruning in game trees reduces the number of nodes evaluated in decision-making processes.

Combinatorial Properties

Cayley's Formula

Cayley's formula states thfor number of labeled trees possible with

n $n-2$. This formula is fundamental in enumerative combinatorics and havepplications in understanding molecule structures in chemistry.

Prufer Code

Prufer code provides a unique representation for labeled trees. This bijective encoding simplifies operations like tree enumeration and random tree generation.

Kirchhoff's Matrix Tree Theorem

This theorem provides a way to count the number of spanning trees in the Graph using the determinant of its Laplacian matrix. It havepplications in electrical networks and combinatorial optimization.

Advanced Topics in Tree Theory

Tree theory is a fundamental branch of graph theory that studies trees-connected, acyclic graphs. Trees serve as a versatile tool in diverse applications, from computer science and biology to network design and optimization. Advanced topics in tree theory delve deeper inforir structural, combinatorial, and algorithmic properties, exploring the powerful insights they provide in both theoretical and practical contexts.

Types of Trees

Spanning Trees

A spanning tree of a graph is a subgraph that connects all vertices of the original graph without forming any cycles. Every connected graph havet least one spanning tree. Key properties include:

Minimum Spanning Tree (MST): A spanning tree with the smallest possible total edge weight. Algorithms such as Kruskal's and Prim's are widely used to find MSTs.

Maximum Spanning Tree: A spanning tree with the largest possible total edge weight, useful in applications like clustering.

Binary Trees

Binary trees are trees where each vertex havet most two children. They are central in computer science applications, particularly in search and sorting algorithms, and can be further classified into:

Full Binary Trees: Each vertex has 0 or 2 children.

Complete Binary Trees: All levels except the last are fully filled, and the last level havell nodes as far left as possible.

Balanced Binary Trees: Heights of the left and right subtrees differ by at most one at every node, ensuring efficiency in operations.

Rooted Trees

In rooted trees, one vertex is designated as a root, and all edges are directed away from or toward it. Rooted trees are used in hierarchical data structures and representations like file systems or organizational charts.

Steiner Trees

Steiner trees are used in network design, where the goal is to connect a subset of vertices (called terminals) with the minimum total edge weight. This problem generalizes the spanning tree problem and is NP-hard.

Decision Trees

Decision trees are rooted trees used in machine learning for classification and regression tasks. They provide an intuitive way to model decision-making processes.

Structural Properties of Trees

Diameter and Radius

Diameter: The longest path between any two vertices in a tree. It is used in network analysis to evaluate the efficiency of communication.

Radius: The minimum eccentricity (maximum distance from a node to any other node) of all vertices in the tree.

Tree Decomposition

Tree decomposition breaks a graph into a tree-like structure, where each node of the tree represents a subset of vertices of the graph. It is particularly useful in algorithms for solving NP-hard problems on graphs with bounded treewidth.

Treewidth

Treewidth measures how close a graph is to being a tree. Graphs with small treewidth allow for efficient solutions to problems like vertex coloring and constraint satisfaction using dynamic programming.

Algorithms in Tree Theory

Tree Traversal Algorithms

Traversal algorithms are fundamental in tree theory, with various techniques tailored for specific applications:

Depth-First Search (DFS): Explores as far as possible along each branch before backtracking. DFS is used in cycle detection, topological sorting, and game search trees.

Breadth-First Search (BFS): Explores neighbors' level by level, ideal for finding the shortest path in unweighted graphs.

Inorder, Preorder, and Postorder Traversals: Specific to rooted binary trees, these traversals have applications in expression tree evaluations, XML parsing, and more.

Tree Center and Centroid

Tree Center: Vertices that minimize the maximum distance to other vertices in the tree. Centers are crucial for designing efficient networks and communication systems.

Tree Centroid: Vertices that minimize the sum of distances to all other vertices. Centroids are used in hierarchical clustering and data analysis.

Tree Pruning

Tree pruning involves removing unnecessary nodes or branches to optimize performance or simplify structure. For example, alpha-beta pruning in game trees reduces the number of nodes evaluated in decision-making processes.

Combinatorial Properties

Cayley's Formula

Cayley's formula states thfor number of labeled trees possible with

n^{n-2}. This formula is fundamental in enumerative combinatorics and havepplications in understanding molecule structures in chemistry.

Prufer Code

Prufer code provides a unique representation for labeled trees. This bijective encoding simplifies operations like tree enumeration and random tree generation.

4.3 Kirchhoff's Matrix Tree Theorem

This theorem provides a way to count the number of spanning trees in the Graph using the determinant of its Laplacian matrix. It havepplications in electrical networks and combinatorial optimization.

CHAPTER 11

APPLICATIONS OF GRAPH THEORY

11.1 Network and Communication Analysis in Graph Theory

Graph theory plays a central role in the analysis and optimization of networks and communication systems. By representing networks as graphs, we can study and improve complex, interconnected systems such as a internet, telecommunications, and transportation networks. This section explores how graph theory applies to network and communication analysis, discussing key concepts such as **network flow**, **connectivity**, and **shortest paths**. Through practical examples, we see how graph theory helps optimize data flow, enhance network reliability, and support critical infrastructure.

1. Modeling Complex Networks with Graphs

In network and communication analysis, a network is often represented as a graph, where nodes (vertices) represent points in the network-such as computers, routers, or locations-and edges represent connections or communication paths between them. This abstraction allows complex systems to be analyzed and optimized through various graph-theoretic concepts.

Types of Graphs in Network Analysis:

1.   **Undirected Graphs:** In undirected networks, such as peer-to-peer or mesh networks, connections are bidirectional. An undirected edge between nodes A and B indicates that data can flow both ways.

2.   **Directed Graphs:** Many communication systems, like data flow in the internet, require directionality. In directed graphs, each edge have direction, meaning data can flow from node A to node B, but not necessarily the other way.

3.   **Weighted Graphs:** In transportation or telecommunication networks, edges may have weights that represent distance, bandwidth, or cost. Weighted graphs allow for optimization of paths based on these criteria.

Example - The Internet as a Graph: The internet can be visualized as a massive, global graph where each device (computer, router) is a node, and connections (physical cables or wireless links) are edges. This representation is crucial for understanding data flow, identifying bottlenecks, and ensuring efficient communication.

2. Network Flow Analysis

Network flow is an essential concept in graph theory applied to communication networks. In a flow network, each edge have capacity representing the maximum amount of data or traffic it can handle, and flow represents the current amount being transferred.

Max-Flow Problem: One of the core problems in network analysis is the **maximum flow problem**, which seeks to determine the maximum feasible flow from a source node SSS to a destination node TTT in a network. Solutions to this problem help optimize data transfer and are applicable in load balancing and traffic management.

Algorithms for Network Flow:

1. **Ford-Fulkerson Algorithm:** This algorithm finds the maximum flow by iteratively identifying augmenting paths with available capacity and increasing the flow until no more augmenting paths exist.

2. **Edmonds-Karp Algorithm:** A specific implementation of the Ford-Fulkerson method that uses breadth-first search (BFS) to find augmenting paths, achieving a more predictable runtime in practical applications.

Example - Optimizing Data Transfer in a Communication Network: Consider a network of routers connected by data cables, where each cable have maximum bandwidth (capacity). By modeling this network as a flow graph, administrators can use the Ford-Fulkerson algorithm to identify the maximum amount of data that can be transferred from one central server to a client network, thereby optimizing usage and avoiding overload.

3. Connectivity and Reliability

Connectivity is a crucial aspect of network analysis, as it determines the robustness and reliability of the network. In graph theory, connectivity is measured by the minimum number of nodes or edges that must be removed to disconnect the network. Highly connected networks are more resilient to failures, making connectivity a critical consideration in designing reliable systems.

Key Connectivity Concepts:

1. **Vertex Connectivity:** This measures the minimum number of nodes (vertices) that need to be removed to disconnect the network. High vertex connectivity implies a network is more robust against node failures.

2. **Edge Connectivity:** This measures the minimum number of edges that need to be removed to separate the network. Networks with high edge connectivity are less susceptible to connection disruptions.

Example - Ensuring Robust Communication in Telecommunications: Telecommunication providers use connectivity analysis to determine the vulnerability of their network infrastructure. For instance, they may calculate the vertex and edge connectivity of their network to ensure that removing a small number of nodes or connections does not disrupt service, especially in critical areas such as emergency communication channels.

4. Shortest Path Analysis for Efficiency

Finding the shortest path between nodes is fundamental to optimizing networks. The shortest path problem seeks the minimum distance or lowest-cost route between two nodes in a weighted graph. In communication and transportation networks, shortest path algorithms improve routing efficiency and reduce travel times or data transmission delays.

Key Algorithms for Shortest Paths:

1. **Dijkstra's Algorithm:** Used for finding the shortest path from a single source node to all other nodes in the Graph with non-negative weights. It is widely applied in network routing and GPS navigation.

2. **Bellman-Ford Algorithm:** This algorithm also finds the shortest paths but can handle graphs with negative weights, making it useful for specific applications in network analysis.

Example - Routing in Telecommunication Networks: Telecommunication networks rely on shortest path algorithms to route calls, messages, and data packets efficiently. Dijkstra's algorithm helps identify the quickest route from one location to another, ensuring minimal delay and efficient use of network resources. For example, routing data packets through a network of routers to reach an endpoint with the least transmission delay.

5. Practical Applications of Graph Theory in Network and Communication Systems

Graph theory is extensively applied in the real world to address network-related challenges. Below are some notable examples:

A. Load Balancing in Data Centers: Data centers consist of thousands of servers connected in a network. Load balancing ensures that each server handles an appropriate amount of traffic, preventing overloads. By using graph theory to model server connections, administrators can analyze flow and optimize traffic distribution, avoiding bottlenecks and maintaining performance.

B. Route Optimization in Transportation Networks: Transportation networks, such as subway systems and airline routes, benefit from graph theory to optimize routes and reduce travel times. By representing stations or cities as nodes and routes as edges, transportation planners use shortest path algorithms to determine the most efficient travel paths, especially during high-demand periods.

C. Resilience in Power Grids: Power grids are modeled as graphs, where power stations and substations are nodes and transmission lines are edges. Graph theory helps analyze grid connectivity and identify critical nodes whose failure could lead to cascading outages. By ensuring high connectivity, power companies can enhance grid reliability and reduce the risk of blackouts.

D. Internet Routing and Autonomous Systems: The internet is a global network composed of many smaller networks, known as autonomous systems (AS). Each AS operates independently, but they are connected to form the internet. By modeling each AS as a node and connections as edges, network engineers use graph theory to optimize data routing, balancing load across paths and ensuring efficient internet service.

11.2 Biological and Ecosystem Modeling

Graph theory provides invaluable tools for understanding complex biological systems and ecosystems. By representing elements such as species, proteins, or neurons as nodes, and their interactions as edges, graph theory allows researchers to visualize and analyze these intricate networks. This section explores key applications of graph theory in biological and ecological modeling, including protein-protein interaction networks, food webs, and neural networks.

1. Protein-Protein Interaction (PPI) Networks

In biological systems, proteins rarely function in isolation; instead, they interact with each other to perform essential cellular tasks. These interactions are studied using Protein-Protein Interaction (PPI) networks, where nodes represent proteins and edges represent interactions between them.

Modeling PPI Networks:

A PPI network is typically represented as an undirected graph where each node corresponds to a protein and each edge indicates an interaction between two proteins. The presence or absence of edges provides insights into potential functional relationships between proteins, as well as pathways critical for cellular functions.

Applications of PPI Networks:

1.   **Understanding Disease Mechanisms:** Analyzing PPI networks helps in identifying proteins involved in disease pathways. For example, researchers can

target specific proteins that interact closely with others implicated in cancer or neurodegenerative diseases.

2. **Drug Discovery:** By studying PPI networks, researchers can identify potential drug targets. Proteins central to disease pathways can be targeted for therapeutic interventions, making the network structure essential for drug design.

3. **Functional Modules and Pathways:** Clustering techniques applied to PPI networks can identify groups of proteins that work together in cellular processes, providing insights into biological pathways.

Example - Cancer Research Using PPI Networks:

In cancer research, PPI networks have been used to identify proteins that play critical roles in tumor growth. Proteins with high centrality in the network are often key to cell proliferation and survival. For instance, in breast cancer studies, high-centrality nodes in the PPI network can reveal potential biomarkers or therapeutic targets.

2. Food Webs in Ecology

A food web represents the feeding relationships between species in an ecosystem, forming a directed graph where nodes represent species and edges represent the direction of energy flow from prey to predator. By analyzing food webs as directed graphs, ecologists gain insights infor stability and resilience of ecosystems.

Modeling Food Webs:

In a food web, each species is represented as a node, and a directed edge from species A to species B signifies that A is a prey species for predator B. The web structure reflects trophic levels and enables the study of energy transfer within the ecosystem.

Applications of Food Web Analysis:

1. **Ecosystem Stability and Collapse Prediction:** Graph theory helps analyze which species play critical roles in maintaining ecosystem stability. Nodes with high connectivity (keystone species) are essential for ecosystem balance; their removal could lead to collapse.

2. **Understanding Trophic Cascades:** Food webs enable ecologists to study the effects of changes in one species on others in the network. For example, the removal of an apex predator may lead to an increase in herbivore populations, affecting vegetation and ecosystem health.

3. **Conservation Efforts:** Identifying keystone species or vulnerable nodes

helps conservationists prioritize species and habitats for protection, as ase nodes are critical for ecosystem resilience.

Example - Food Web of Marine Ecosystems:

In marine ecosystems, food webs can be highly complex, with various interdependencies among species. For example, in the North Sea, fish such as herring and cod play significant roles in the food web, linking smaller prey like zooplankton to larger predators such as seals. By modeling this system as a food web, scientists can predict the impact of overfishing on the ecosystem, as removing or reducing one species can disrupt the entire food chain.

3. Neural Networks in Neuroscience

The human brain is an intricate network of neurons connected through synapses, forming an extensive neural network. Representing neural connections as a graph enables neuroscientists to study the structural and functional connectivity of the brain, providing insights into cognition, behavior, and disorders.

Modeling Neural Networks:

In a neural network graph, neurons are nodes, and synaptic connections are directed edges, where the direction indicates the flow of electrical impulses between neurons. The graph structure allows for the examination of connectivity patterns, such as clusters of highly interconnected neurons (modules) or critical connections that influence brain activity.

Applications of Neural Network Analysis:

1. **Mapping Brain Connectivity (Connectomics):** Graph theory allows researchers to map neural connections across different regions of the brain. These maps, known as connectomes, help identify brain regions that play central roles in cognition and behavior.

2. **Understanding Neurological Disorders:** Abnormalities in neural networks are linked to various mental health conditions and neurodegenerative diseases. Analyzing the connectivity patterns of patients with conditions like Alzheimer's or schizophrenia can reveal changes in the brain network structure.

3. **Studying Functional Modules:** By identifying clusters of neurons that work together, neuroscientists can understand how different brain regions contribute to specific functions, such as memory, perception, and movement.

Example - Analyzing Alzheimer's Disease with Neural Networks:

In Alzheimer's research, graph theory helps analyze disruptions in brain networks. By comparing healthy and diseased brains, researchers observe that

Alzheimer's patients exhibit reduced connectivity in regions associated with memory and cognition. This disruption manifests as decreased centrality and clustering in specific areas, providing insights infor progression of the disease.

4. Ecosystem Stability and Resilience Analysis

Ecosystem stability is vital for maintaining biodiversity and ecosystem services. Graph theory aids in understanding ecosystem resilience by modeling how species interactions impact ecosystem stability and identifying potential vulnerabilities.

Modeling Ecosystem Stability:

Graphs model ecosystems by representing species as nodes and interactions as edges, allowing ecologists to simulate the effects of various perturbations on the ecosystem. Highly connected nodes (species with numerous interactions) are often critical to maintaining stability, while weakly connected nodes may be less influential.

Applications of Ecosystem Stability Analysis:

1.   **Predicting Ecosystem Responses to Environmental Changes:** By examining how the removal of certain species impacts the network, ecologists can predict ecosystem responses to factors like climate change, pollution, or invasive species.

2.   **Identifying Keystone Species:** Certain species play pivotal roles in maintaining ecosystem structure. Removing these species could cause cascading effects, potentially leading to ecosystem collapse.

3.   **Habitat Fragmentation and Connectivity:** Graphs help model the effects of habitat fragmentation, where species are increasingly isolated. Graph theory can identify critical connections (wildlife corridors) necessary to maintain biodiversity.

Example - Analyzing Rainforest Ecosystems:

In tropical rainforests, many species are interdependent, creating complex food webs. For example, certain plant species depend on specific animals for pollination or seed dispersal. By modeling these dependencies as a network, researchers can identify keystone species that are essential for the health and resilience of the ecosystem. Removing or declining these species, such as key pollinators, would disrupt the network and threaten biodiversity.

5. Microbial Interaction Networks

Microbial communities, such as those in soil, the human gut, or aquatic environments, are composed of diverse species that interact in complex ways.

Graph theory allows microbiologists to study these interactions and understand how microbes contribute to health, environmental processes, and nutrient cycling.

Modeling Microbial Networks:

Microbial interaction networks are represented as graphs where nodes are microbial species, and edges indicate interactions, such as mutualism, competition, or predation. By studying these networks, scientists gain insights infor stability and dynamics of microbial communities.

Applications of Microbial Network Analysis:

1. **Understanding Gut Health:** In human health, gut microbiota play a critical role in digestion, immunity, and disease. Graph theory enables analysis of microbial diversity and interactions, helping to understand the balance of beneficial and harmful microbes.

2. **Environmental Impact Studies:** Microbes contribute to nutrient cycling and pollutant breakdown in ecosystems. By analyzing microbial networks in soil or water, researchers can assess ecosystem health and resilience.

3. **Disease Prevention:** Some microbes inhibit pathogens, while others promote their growth. Graph analysis can help identify beneficial microbial communities that suppress harmful pathogens in agriculture or human health.

Example - Human Gut Microbiome:

The human gut microbiome comprises a vast network of microbes. By analyzing this network, researchers identify clusters of beneficial bacteria that aid digestion and immune function. Changes in this network, such as a loss of microbial diversity, are associated with health conditions like obesity, inflammatory bowel disease, and metabolic disorders.

11.3 Social Network Analysis and Community Detection in Graph Theory

Graph theory has become a cornerstone in social network analysis, providing a mathematical framework for representing and analyzing relationships among individuals or groups. By modeling social networks as graphs, researchers can gain insights into community structures, the influence of individuals, and patterns of group dynamics. This section explores how graphs represent social networks and delves into methods such as **modularity, clustering,** and **betweenness centrality**, which help in detecting communities and understanding social behavior.

1. Representing Social Networks as Graphs

In social network analysis (SNA), individuals or entities are represented as nodes (vertices) and the connections or relationships between them are represented as edges. These connections can signify various forms of interactions, such as friendships, collaborations, or shared interests.

Types of Social Network Graphs:

1. **Undirected Graphs:** These graphs represent mutual relationships. For example, if nodes A and B represent two friends, an undirected edge between A and B signifies a mutual friendship.

2. **Directed Graphs:** Used when relationships are one-way, such as a "follows" relationship on social media. An edge from node A to node B indicates that A follows B, but not necessarily vice versa.

3. **Weighted Graphs:** In weighted social networks, edges have weights that reflect the strength of the relationship, such as a frequency of interactions or the duration of a connection.

Example - Social Media Networks: Social media platforms like Facebook and Twitter can be modeled as graphs. On Facebook, friendships are typically mutual, forming an undirected graph. Twitter's "following" relationships, however, create a directed graph, as users can follow others without reciprocity. Modeling these networks as graphs allows for detailed analysis of user behavior, influence, and community structure.

2. Community Structure and Detection

Communities in social networks are groups of individuals or entities that are more densely connected to each other than for rest of the network. Detecting these communities is crucial for understanding group dynamics, identifying influential members, and predicting behaviors within a network.

Modularity and Community Detection: Modularity is a measure used to assess the strength of division in a network, helping to identify communities by maximizing intra-group connections and minimizing inter-group connections. High modularity values indicate strong community structures, where nodes within a community are more connected to each other than to nodes outside the community.

1. **Modularity Maximization:** The goal is to partition the graph into clusters that maximize modularity. This method identifies communities by maximizing the density of connections within groups compared to between groups.

2. **Louvain Algorithm:** This is a popular modularity-based algorithm that uses a two-phase process to detect communities. In the first phase, nodes are grouped to optimize modularity locally; in the second phase, these groups are

aggregated to form a single node, and the process repeats.

Example - Detecting Communities in Online Social Networks: In online social networks, modularity-based algorithms like the Louvain method are used to detect clusters of users with common interests or behaviors. For instance, in a network like Facebook, these clusters could represent groups of people who frequently interact, share similar posts, or have mutual friends.

3. Centrality Measures for Influence and Importance

Centrality measures are essential for identifying influential nodes within a social network. Each measure captures a different aspect of influence, whether it be the number of connections, the closeness of connections, or the control a node has over interactions between other nodes.

Types of Centrality Measures:

1. **Degree Centrality:** This measures the number of direct connections a node has. Nodes with high degree centrality are often influencers in a network, having many immediate connections.

2. **Closeness Centrality:** This measure is based on the average shortest path from a node to all other nodes in the graph. Nodes with high closeness centrality can quickly interact with all other nodes, making them valuable for spreading information.

3. **Betweenness Centrality:** This metric measures the extent to which a node lies on the shortest paths between other nodes. Nodes with high betweenness centrality act as "bridges" within the network, controlling the flow of information between different parts of the network.

Example - Identifying Influencers in Social Media: In social media networks, centrality measures help identify influential users. For instance, on Twitter, users with high degree centrality have many followers, while those with high betweenness centrality are more likely to bridge communities and spread information across diverse groups.

4. Clustering in Social Network Analysis

Clustering is a fundamental approach in SNA to group similar nodes together based on their connections. Clustering not only identifies communities but also reveals the overall structure and organization of a social network.

Clustering Coefficient: The clustering coefficient of a node indicates how tightly connected its neighbors are. A high clustering coefficient suggests thfor neighbors of a node are likely interconnected, forming local clusters or communities. Clustering is particularly valuable for understanding group cohesion within social networks.

Hierarchical Clustering: Hierarchical clustering organizes nodes into a tree-like structure, where nodes with strong connections are placed together for bottom levels, and broader groups emerge at higher levels. This method is beneficial for identifying nested communities, where smaller groups are part of larger ones.

Example - Analyzing Collaboration Networks: In scientific research networks, hierarchical clustering can reveal how researchers collaborate within and across disciplines. For example, within a network of researchers, smaller clusters may represent collaborations on specific projects, while higher-level clusters might indicate collaborations within broader fields of study.

5. Applications of Community Detection in Social Networks

Community detection provides valuable insights infor structure and behavior of social networks, supporting applications in various fields, including marketing, public health, and political analysis.

A. Marketing and Targeted Advertising: Community detection is widely used in marketing to identify clusters of users with similar interests or behaviors. By targeting these clusters, companies can deliver personalized ads to relevant audiences, improving engagement and conversion rates.

Example - Targeting User Communities on Social Media: On platforms like Instagram, brands can use community detection algorithms to identify groups of users interested in specific products or trends. For example, a fitness brand might target a community of users who follow health and wellness influencers, maximizing the impact of their marketing campaigns.

B. Public Health and Disease Spread Analysis: Graph theory and community detection help public health officials understand how diseases spread within social networks. By identifying communities and high-centrality individuals, health interventions can be more effectively targeted to contain outbreaks.

Example - Tracing Disease Spread in Communities: In the case of infectious diseases like COVID-19, health authorities used community detection and centrality measures to track the spread of the virus within social networks. Identifying clusters of highly connected individuals helped prioritize testing and vaccination efforts, reducing further spread.

C. Political and Social Movement Analysis: Community detection also plays a role in political science and sociology by identifying influential groups within social movements. Analyzing the central nodes in these networks reveals individuals or organizations that drive the movement's momentum.

Example - Analyzing Social Movements on Twitter: On platforms like Twitter, social movements often form distinct communities. For instance, during political campaigns, central nodes in these networks might represent key influencers, while communities reflect voter groups or advocacy clusters. By detecting these groups, political analysts can gauge the movement's reach and influence.

11.4 Computing and Databases

Graph theory provides powerful tools for solving complex problems in computing and database systems. Its versatility allows for efficient data organization, analysis, and management across various applications, including graph databases, dependency graphs, and resource management. This section explores how graph theory enhances computing and database efficiency, with specific use cases in query optimization, resource scheduling, and dependency resolution.

1. Graph Databases and Query Optimization

Traditional databases, which use tables and rows, can struggle with complex relationships between data elements. Graph databases, however, are designed specifically to store and analyze relationships, making them ideal for highly interconnected data. In the Graph database, data is stored as nodes, edges, and properties, closely aligning with graph theory's structure.

Modeling Data with Graph Databases:

In the Graph database, each entity is represented as a node, and relationships between entities are represented as edges. This structure allows for the direct storage of relationships, reducing the need for complex joins and enabling fast queries on relationships and paths.

Key Features of Graph Databases:

1. **Efficiency in Handling Relationships:** Graph databases excel in scenarios where data is heavily interconnected, such as social networks, recommendation systems, and fraud detection.

2. **Fast Querying with Graph Traversals:** Instead of using table joins, graph databases employ traversals, which allow queries to move from one node to connected nodes efficiently.

Example - Query Optimization in Social Networks:

Consider a social media platform where users are connected as friends. In a relational database, finding mutual friends between two users would require complex join operations, especially with large datasets. In the Graph database, however, this query becomes straightforward: by starting at one user's node and traversing connections, the database can quickly identify mutual friends. Graph databases like Neo4j use optimized graph traversal algorithms to speed up such queries, allowing real-time responses.

2. Dependency Graphs in Software Development and Build Systems

Dependency graphs are used extensively in software development to manage and understand dependencies between components, libraries, and systems. By representing dependencies as directed graphs, where nodes are modules and edges indicate dependency relationships, developers can track dependencies and manage build processes more efficiently.

Applications of Dependency Graphs:

1. **Build Systems:** In large projects, build systems use dependency graphs to determine the order in which components need to be compiled. If component A depends on component B, then B must be compiled before A.

2. **Library Management:** Dependency graphs help manage libraries in a project, especially when updating or replacing libraries. The graph structure enables developers to assess the impact of changes on other parts of the system.

Example - Dependency Resolution in Build Systems:

In software like Make or Gradle, dependency graphs optimize the build process by identifying components that can be built in parallel, reducing compilation time. In a complex project with numerous dependencies, such as an enterprise application, these tools use dependency graphs to build a directed acyclic graph (DAG) of tasks, ensuring that dependencies are met while minimizing redundant builds.

3. Resource Management and Scheduling

In computing environments like cloud computing, data centers, and operating systems, resource management involves allocating and scheduling resources such as CPU time, memory, and storage. Graph theory supports resource management by modeling tasks, resources, and dependencies as graphs, optimizing the allocation and scheduling processes.

Graph Theory in Resource Scheduling:

1. **Task Scheduling in Operating Systems:** Operating systems use scheduling algorithms to assign CPU time to tasks. Dependency graphs help manage the execution order, especially for tasks with dependencies.

2. **Job Scheduling in Data Centers:** In cloud and distributed computing, tasks are scheduled across multiple servers. By using directed graphs to model task dependencies and resource requirements, scheduling algorithms can optimize resource usage and minimize delays.

Example - Task Scheduling in Parallel Computing:

In parallel computing, tasks often have dependencies, meaning some tasks cannot start until others finish. Dependency graphs model these dependencies, enabling the scheduler to determine which tasks can run concurrently. Graph-based scheduling algorithms, like critical path scheduling, help identify bottlenecks and prioritize tasks to maximize resource utilization and minimize total execution time.

4. Graph-Based Data Models in Data Analysis and Machine Learning

Graph theory enhances data models used in data analysis and machine learning, particularly when dealing with complex relationships. Graph-based data models can represent networks of entities with intricate relationships, making them ideal for recommendation systems, fraud detection, and clustering.

Graph Neural Networks (GNNs):

In machine learning, Graph Neural Networks (GNNs) are specialized neural networks that operate on graph-structured data. GNNs aggregate information from a node's neighbors, allowing them to capture relational information that traditional neural networks might overlook.

Applications of GNNs:

1. **Recommendation Systems:** By modeling user-product interactions as a graph, recommendation algorithms can identify similar users or products based on shared connections.

2. **Fraud Detection:** Graphs are used in financial networks to detect fraud by identifying unusual patterns. For instance, a node with an unusual number of connections or certain patterns of interactions can signify fraudulent behavior.

Example - Graph-Based Recommendations on E-Commerce Platforms:

E-commerce platforms, like Amazon, use graph models to recommend products. By constructing a graph where users and products are nodes and interactions (e.g., purchases, clicks) are edges, recommendation algorithms can identify products similar to those a user has interacted with, increasing the accuracy and relevance of recommendations.

5. Network Analysis for Dependency and Impact Evaluation

Graph theory supports dependency and impact evaluation in network analysis by modeling components and their relationships. These analyses are essential in complex systems like telecommunications, where understanding dependencies helps prevent cascading failures.

Network Impact Analysis:

By modeling a system as a dependency graph, impact analysis can determine how the failure or disruption of one component affects others. Centrality measures, such as betweenness and closeness centrality, can identify critical nodes whose failure would disrupt the network.

Example - Analyzing IT Infrastructure with Dependency Graphs:

In IT infrastructure, dependency graphs represent systems and their connections, such as servers, databases, and applications. By simulating failures within this graph, administrators can identify vulnerable points, helping them develop strategies for redundancy and resilience.

6. Optimization in Graph Databases

Graph databases are optimized for handling vast networks and relationships, offering efficient data retrieval methods that are crucial for recommendation engines, social networks, and transportation systems. Their primary advantage lies in the ability to store and query relationships without needing complex joins, making them faster and more flexible than traditional relational databases.

Graph Algorithms for Optimization:

1. **PageRank and Influence:** Algorithms like PageRank, originally used by Google to rank web pages, are also used in graph databases to determine the importance of nodes based on connections. This is useful in social media and e-commerce recommendations.

2. **Shortest Path Queries:** Graph databases allow for efficient shortest path computations, which are essential for applications like navigation and route planning.

Example - Graph-Based Optimization in Transportation Systems:

In transportation, where route optimization is critical, graph databases help identify the fastest or most cost-effective routes. By using shortest-path algorithms like Dijkstra's or A*, transportation companies can optimize delivery routes or reduce travel time, resulting in lower costs and improved service.

7. Security and Access Control Graphs

Graph theory also applies to security by representing access permissions and connections as graphs. In this context, nodes represent users or systems, and edges represent permissions or interactions.

Graph-Based Security Applications:

1. **Access Control:** Access control lists (ACLs) and role-based access control (RBAC) systems can be represented as graphs, enabling administrators to visualize and manage permissions.

2. **Intrusion Detection:** Network security systems model connections between users, devices, and applications as a graph to detect unusual patterns or anomalies, which could indicate security threats.

Example - Access Control in Enterprise Environments:

In an enterprise setting, employees' access to systems and resources is often managed through RBAC, where roles are nodes and permissions are edges. Graphs make it easy to visualize and analyze access hierarchies, ensuring that sensitive data is protected while employees can access the resources they need.

11.5 Future Directions and Research Opportunities in Graph Theory

Graph theory continues to be a dynamic area of research, driven by its vast applications in diverse fields. As technology and data complexity grow, new avenues in graph theory are emerging, promising to address complex problems in innovative ways. This section explores key research opportunities and future directions in graph theory, focusing on **dynamic graph analysis**, **quantum computing applications**, and **graph theory's expanding role in artificial intelligence**. It also highlights recent advancements and open questions that researchers aim to tackle.

1. Dynamic Graph Analysis

Traditionally, graph theory has focused on static graphs, where nodes and edges remain constant. However, many real-world networks-such as social networks, communication networks, and biological networks-are dynamic, with changing structures over time. **Dynamic graph analysis** has emerged to address this variability, offering ways to model and analyze time-evolving data.

Research Goals in Dynamic Graph Analysis:

1. **Modeling Time-Dependent Changes:** Dynamic graphs, also called temporal graphs, track changes in edges and nodes over time. This approach is essential for understanding the temporal patterns within social networks, transportation systems, and epidemiological models.

2. **Algorithm Development for Real-Time Analysis:** Dynamic graphs require algorithms that can adjust to changes without recalculating the entire graph structure, enabling real-time analysis of evolving systems.

3. **Applications in Anomaly Detection:** In network security, dynamic

graphs can be used to detect anomalies, such as unusual spikes in data flow, by observing changes over time. This has significant implications for detecting cyber threats and fraud.

Recent Advances and Open Questions: Recent work in dynamic graph analysis includes developing algorithms for real-time updates, as well as techniques for efficiently storing and processing time-evolving graphs. Open questions in this field focus on scalability-how to handle large, dynamic networks with minimal computational resources-and on the robustness of these models in capturing meaningful temporal patterns.

2. Quantum Computing Applications in Graph Theory

Quantum computing represents a significant shift in computational power, and its potential impact on graph theory is profound. Graph-related problems, especially those involving optimization and search, often suffer from high computational complexity. Quantum algorithms, with their ability to handle superpositions and entanglements, offer the possibility of solving some of these problems exponentially faster than classical algorithms.

Potential Areas of Impact:

1. **Quantum Algorithms for Graph Problems:** Quantum algorithms like **Grover's search** and **Shor's algorithm** have already demonstrated exponential speedup for certain types of problems. Researchers are now exploring quantum algorithms specifically designed for graph problems, such as graph isomorphism and pathfinding.

2. **Quantum Graph States:** Quantum states can represent graph structures in the way that enables faster processing and parallelism. **Quantum graph states** are being studied for applications in quantum communication and cryptography.

3. **Optimization in Quantum Networks:** Problems such as finding the shortest path, maximum flow, and minimum spanning trees are integral to network optimization. Quantum computing's potential for solving these optimization problems efficiently has implications for areas like logistics, data routing, and financial modeling.

Example - Quantum Walks for Faster Graph Traversal: Quantum walks, a quantum equivalent of random walks, are a promising method for faster graph traversal. For instance, quantum walks can enhance algorithms for page ranking and social network analysis by quickly identifying influential nodes. Researchers are exploring how quantum walks can improve traversal times and reduce computational demands for large-scale networks.

3. Artificial Intelligence and Machine Learning with Graph Theory

Graph theory is increasingly integrated into **artificial intelligence (AI)** and **machine learning (ML)**, especially for tasks requiring relational understanding and pattern recognition. AI models now incorporate graph structures to understand complex, interconnected data, with applications in natural language processing, social network analysis, and recommendation systems.

Key Research Areas in AI and Graph Theory:

1. **Graph Neural Networks (GNNs):** GNNs are a new class of neural networks that leverage graph structures, allowing AI models to learn from relational data. This approach has transformed fields such as social network analysis, drug discovery, and knowledge graph construction.

2. **Graph-Based Semi-Supervised Learning:** Many machine learning tasks have limited labeled data. Semi-supervised learning on graphs uses graph structures to propagate labels from labeled nodes to unlabeled ones, improving learning accuracy without extensive labeled data.

3. **Explainability in AI:** As AI systems become more complex, explaining their decisions becomes crucial. Graph-based approaches can help by mapping decision paths and identifying key influences, enhancing transparency and trust in AI.

Recent Advancements and Open Challenges: The rise of GNNs has led to breakthroughs in various domains, from predicting protein interactions to recommending content in social networks. However, challenges remain in scaling GNNs to massive networks, addressing data sparsity, and improving model interpretability. Researchers are also exploring how graph-based models can improve fairness and reduce biases in AI systems.

4. Applications of Graph Theory in Cybersecurity

Graph theory offers tools for identifying and analyzing relationships within cybersecurity frameworks, from detecting vulnerabilities to analyzing malware networks. As cyber threats evolve, graph theory's role in cybersecurity continues to grow, particularly in the area of network security.

Cybersecurity Applications of Graph Theory:

1. **Anomaly and Intrusion Detection:** Network traffic can be modeled as a graph, with nodes representing devices and edges representing communications. By monitoring for unusual patterns, graph-based models can detect potential security breaches.

2. **Threat Intelligence Graphs:** Graphs enable visualization of complex relationships between threats, vulnerabilities, and exploits, assisting security teams in tracing attack origins and anticipating potential threats.

3. **Malware Analysis and Detection:** Graph-based analysis of malware binaries can reveal similarities between malware strains, enabling faster and more accurate detection of threats based on known attack patterns.

Open Questions and Research Focus: The complexity of modern networks poses scalability challenges for graph-based cybersecurity solutions. Researchers are working on distributed graph algorithms and machine learning integration to improve response times and accuracy in detecting threats across expansive networks.

5. Graph Theory in Social and Economic Networks

Social and economic networks are some of the most complex and dynamic systems that researchers seek to understand. Graph theory's ability to model and analyze relationships makes it invaluable for studying phenomena like social influence, market dynamics, and information diffusion.

Research Areas in Social and Economic Graphs:

1. **Influence Propagation Models:** In social networks, understanding how information or behaviors spread is crucial for predicting trends and designing effective marketing strategies. Graph theory helps model these dynamics by analyzing how influence flows through network structures.

2. **Resilience in Economic Networks:** Economic systems, including trade networks and financial markets, are interconnected, with disruptions in one part of the network often causing ripple effects. Graph-based resilience analysis helps identify critical nodes and assess the robustness of economic systems.

3. **Game Theory on Graphs:** In economic and social networks, game theory on graphs models interactions among agents, capturing strategic behaviors in various contexts, such as competition, cooperation, and negotiation.

Example - Influence Maximization in Social Media Campaigns: Influence maximization, a problem where the goal is to select a subset of influential individuals in a network to maximize information spread, is heavily studied in social media marketing. Researchers use graph theory to develop algorithms that identify the optimal set of users, enabling efficient and cost-effective marketing campaigns.

6. Challenges and Open Problems in Graph Theory

Despite recent advances, several open questions in graph theory present exciting research challenges for the future:

A. Graph Isomorphism Problem: Determining whether two graphs are isomorphic-meaning they contain identical structures-is computationally challenging, particularly for large graphs. Although recent breakthroughs, such as László Babai's quasi-polynomial time algorithm, have improved our understanding, an efficient, scalable solution remains elusive.

B. Scalability in Large-Scale Graph Analysis: As graph datasets grow exponentially, there is a pressing need for scalable algorithms that can handle massive graphs efficiently. This is particularly challenging for real-time applications in dynamic networks, like social media or financial systems, where the graph structure changes constantly.

C. Privacy Preservation in Graph-Based Models: With the rise of graph-based models in social media and other domains, preserving privacy is an important consideration. Researchers are exploring methods for anonymizing graph data, but balancing privacy with data utility remains an ongoing challenge.

D. Graph Theory and Quantum Algorithms: Quantum computing's theoretical potential for solving graph-related problems has sparked interest in developing quantum algorithms for various graph theory applications. However, designing quantum-friendly algorithms that leverage entanglement and superposition while handling real-world graph data remains challenging.

9 789348 037237